Optical Spectra of Transparent Rare Earth Compounds

Optical Spectra of Transparent Rare Earth Compounds

S. HÜFNER

Universität des Saarlandes
Saarbrücken, West Germany

ACADEMIC PRESS New York San Francisco London 1978

A Subsidiary of Harcourt Brace Jovanovich, Publishers

ACADEMIC PRESS, INC.
111 Fifth Avenue, New York, New York 10003

United Kingdom Edition published by
ACADEMIC PRESS, INC. (LONDON) LTD.
24/28 Oval Road, London NW1 7DX

Library of Congress Cataloging in Publication Data

Hufner, S.
Optical spectra of transparent rare earth compounds.

Includes bibliographical references.
1. Rare earth metal compounds--Spectra. I. Title.
QC462.R2H82 546'.4 77-74027
ISBN 0-12-360450-8

PRINTED IN THE UNITED STATES OF AMERICA

Contents

Chapter 3 **Trivalent Ions in the Static Crystal Field**

Chapter 4 **Trivalent Rare Earth Ions in a Phonon Field**

Chapter 5 **Energy Transfer**

Chapter 6 **Magnetic Interactions**

Preface

The first spectra of rare earth compounds were observed in 1908 by the French scientist J. Becquerel [Be08]. They showed the feature that has continued to attract solid state physicists: an extreme sharpness of the lines that is otherwise generally observable only in the spectra of free atoms and molecules. This observation led researchers to hope that by this quality the rare earth crystal spectra would reveal as much about solids as the free ion spectra have disclosed about the atom.

The basic features of rare earth spectra in transparent solids were understood more than a decade ago. Since then many details have been investigated and new applications such as the investigation of magnetism or the Jahn–Teller effect have been found. This field has been so thoroughly researched that it seemed worthwhile to present the most important material in a monograph.

This book is written primarily for people who need an overall view of the field, such as new workers. The treatment is somewhat basic, and emphasis is placed on explaining the underlying physics in selected examples rather than on covering all available material. The book has been written from an experimentalist's point of view. Theoretical results are usually presented without proof in a form that allows their application to the interpretation of experimental data.

The literature in this field is extensive. No attempt was made to collect and present a complete list of references. We have relied on whatever adequate reviews or books exist, and further references may be found in them. We therefore apologize to the many authors whose work is not referenced explicitly. In trying to write an introductory treatment, we thought this procedure appropriate.

In Chapter 1 we present an outline of the book. In Chapters 2 and 3 the properties of free ions and of ions in the crystal field are treated. Subsequent chapters deal with the electron–phonon interaction, energy transfer, magnetic interactions, hyperfine interactions, the Jahn–Teller effect, divalent rare earth ions, glasses, and lasers.

It is certainly appropriate to mention a number of reviews that deal, although from different aspects, with the subject of this monograph: Fick and Joos [FiJo57], McClure [McCl59], Judd [Ju63], Wybourne [Wy65], Dieke [Di68], Crosswhite and Moss [CrMo67], Taylor [Ta72], Di Bartolo [DiBa68, DiBa75] and Jørgensen and Reisfeld [JoRei77]. The author has also benefited very much from the study of the monograph on paramagnetic resonance by Abragam and Bleaney [AbBl70].

Acknowledgments

I thank those numerous authors who allowed me to include material from their work. I am grateful to Professor Hellwege (Darmstadt) in whose laboratory I performed my first work on rare earth crystal spectra. A special word of thanks goes to Professor Kahle (Karlsruhe). He patiently taught me the principles of rare earth physics and also carefully read the entire manuscript, finding errors and making many valuable suggestions. I thank Professor Pelzl (Bochum), who proofread the manuscript, and Dr. Courths (Saarbrücken), who proofread some chapters. I am grateful to Mrs. A. Faltenbacher, who typed most parts of the manuscript. We thank Professor Wachter for supplying us with a valuable preprint [Wa77]: Fig. 43, Table 41; and Drs. Carnall, H. Crosswhite, and H. M. Crosswhite for sending us a very useful report: Tables 4b and 4c.

A final special thanks goes to the publisher for the encouragement, patience, and help that improved the manuscript considerably.

Optical Spectra of Transparent Rare Earth Compounds

CHAPTER 1

Survey

1.1 Introduction

The most impressive feature about the spectra of rare earth (RE) ions in ionic crystals is the sharpness of many lines in the absorption and emission spectra. As early as 1908 [Be08] it was realized that in many cases these lines can be as narrow as those commonly observed in the spectra of free atoms or free molecules. This means they sometimes have a width of not more than 0.01 Å.† For those who think of spectra of solids in terms of very broad absorption lines or bands, this feature must seem particularly attractive; it means that in principle it is possible to investigate interactions in a solid by optical means with a degree of accuracy similar to that usually possible with free atoms or ions.

The goal of the present book is twofold: to explain how these very narrow lines arise and what interactions determine their energy, and to indicate how they can be used to investigate particular properties of the solid state.

The RE ions in ionic crystals have been noticeably successful in technological applications, for example, the red fluorescence of europium

† The narrowest linewidth obtained so far in an optical spectrum of an RE solid are the 200 kHz (6.6×10^{-6} cm^{-1}) in the $^1D_2 \rightarrow {}^3H_4$ transition in $Pr^{3+}:LaF_3$ measured by Erickson [Er 77] with enhanced and saturated absorption spectroscopy.

TABLE 0

THE RARE EARTH ELEMENTS AND SOME OF THEIR BASIC PROPERTIES[a,b]

Atomic number	Element	Electron configuration RE^{3+}	Ground term RE^{3+}	Landé factor g_J RE^{3+}	Max ESR g factor $2M_J(\max)\cdot g_J$	Ionic radius RE^{3+} (Å)	Electron configuration RE^{2+}	Ground term RE^{2+}
57	Lanthanum	$4f^0 5s^2 5p^6$	$^1S_0$	0	0	1.15		
58	Cerium	$4f^1 5s^2 5p^6$	$^2F_{5/2}$	$\frac{6}{7}$	4.29	1.02		
59	Praseodym	$4f^2 5s^2 5p^6$	$^3H_4$	$\frac{4}{5}$	6.40	1.00		
60	Neodymium	$4f^3 5s^2 5p^6$	$^4I_{9/2}$	$\frac{8}{11}$	6.55	0.99		
61	Promethium	$4f^4 5s^2 5p^6$	$^5I_4$	$\frac{3}{5}$	4.80	0.98		
62	Samarium	$4f^5 5s^2 5p^6$	$^6H_{5/2}$	$\frac{2}{7}$	1.43	0.97	$4f^6 5s^2 5p^6$	$^7F_0$
63	Europium	$4f^6 5s^2 5p^6$	$^7F_0$	0	0	0.97	$4f^7 5s^2 5p^6$	$^8S_{7/2}$
64	Gadolinium	$4f^7 5s^2 5p^6$	$^8S_{7/2}$	2	14	0.97		
65	Terbium	$4f^8 5s^2 5p^6$	$^7F_6$	$\frac{3}{2}$	18	1.00		
66	Dysprosium	$4f^9 5s^2 5p^6$	$^6H_{15/2}$	$\frac{4}{3}$	20.00	0.99		
67	Holmium	$4f^{10} 5s^2 5p^6$	$^5I_8$	$\frac{5}{4}$	20.00	0.97		
68	Erbium	$4f^{11} 5s^2 5p^6$	$^4I_{15/2}$	$\frac{6}{5}$	18.00	0.96		
69	Thulium	$4f^{12} 5s^2 5p^6$	$^3H_6$	$\frac{7}{6}$	14.00	0.95		
70	Ytterbium	$4f^{13} 5s^2 5p^6$	$^2F_{7/2}$	$\frac{8}{7}$	8.00	0.94	$4f^{14} 5s^2 5p^6$	$^1S_0$
71	Lutecium	$4f^{14} 5s^2 5p^6$	$^1S_0$	0	0	0.93		

[a] The electron configurations give only the relevant outer electrons. The ground term classification is given in Russell–Saunders symbols. The maximum g-factor as defined by electron spin resonance (ESR) notation is included, because it is frequently found in the literature. Divalent electron configurations are given only where they occur in this volume. The ionic radii are from Evans [Ev64].

[b] Conversion of the frequently used "energy unit" cm^{-1} into more common units: $1\ cm^{-1} \triangleq 2.998 \times 10^{10}$ (1/sec) $\triangleq 1.439$ (K) $\triangleq 1.240 \times 10^{-4}$ (eV).

used for color television screens and the 1.06 μm laser line of neodymium used in various hosts, even glasses.

The narrow optical lines suggest that the interaction or RE ions with the crystalline environment is relatively weak. One can thus describe RE energy levels to a good degree of accuracy with a one-ion model—a very attractive feature in solid state physics.†

Almost all the sharp lines are transitions within the manifold of the 4f electrons, and if not otherwise explicitly stated, those of the 4f electrons will be meant, when we speak of energy levels or optical transitions. We list in Table 0 some basic information about the RE: atomic number, elements, outer electronic shells, free ion ground terms,‡ and Landé g_J-factors for the trivalent state. For convenience we list the so-called maximum electron spin resonance g-factor, which is twice the product of the Landé g_J-factor times the maximum M_J of the ground term, and the ionic radii for the trivalent ions. For divalent ions, electron configurations and ground states are given only for Sm, Eu, and Yb, the most commonly occurring elements in this state. It has become customary in spectroscopy to give the energy in wave numbers or cm^{-1}—a unit that, in order to be a "true energy unit," has to be multiplied by hc (h is Planck's constant and c the velocity of light). The conversion of this unit to some more common ones is given in Table 0, footnote *b*.

1.2 Free Ions

As mentioned many of the absorption and emission lines in the spectra of RE solids are as sharp as lines observed in gas spectra. Since absorption spectra of solids usually show very broad bands, one intuitively guesses

† Recently, however, Cone and Meltzer [CoMe75, Me76] have found energy dispersions as large as 10 cm^{-1} in some levels of $Tb(OH)_3$ (see Chapter 6).

‡ Here a word on the notation is useful. In Russell–Saunders notation the states of a free atom or ion are characterized by a set of quantum numbers S, L, J; such a state is often called a term (see, e.g., [AlFi75, Vol. III, p. 165]). This use of the word term is not to be confused with its use for a part of an equation.

States (terms) with the same S, L but different J are called a multiplet.

In a crystal a free ion state (term) of an RE ion is split into a number of crystal field states. All the crystal field states into which a free ion state is split are often called a crystal field multiplet. In the context of this book the notation ground state can have two meanings: it can be the lowest energy level of a free ion or that of an ion in a crystal; we shall call the former the free ion ground state or ground term and the second crystal field ground state except where the meaning of the word is unambiguous.

that the RE spectra in solids originate from transitions between states that are themselves still quite atomiclike. To understand the energy levels of RE ions in a crystal it is necessary to determine them in detail in the free atom. Since the lines originate from transitions within the 4f manifold (intra-4f transitions—contrary to what is usually observed in atomic spectra), only the energy levels of the 4f electron shell have to be obtained. The energy levels of the 4f electrons in the free atom or ion can be determined with good accuracy from a Hamiltonian which contains their mutual Coulomb repulsion and the spin–orbit interaction. For the 4f electrons these two interaction energies are of roughly the same order of magnitude. This is the intermediate coupling case. To calculate the free ion energy levels from first principles, one has to solve the eigenvalue problem for the Coulomb and spin–orbit interaction simultaneously in order to account for the actual situation. In these calculations the absolute energies are determined by three Coulomb integrals and by the spin–orbit coupling constant. In principle, these constants can be determined from an unrestricted Hartree–Fock calculation [FrWa62]. A comparison with the experimental values obtained shows a calculation accuracy of about 20%.

The generally employed procedure, therefore, is to fit the four energy constants as free parameters to the actually observed energy levels in order to obtain a satisfactory description of the situation. The free ion energy levels of all RE elements have been obtained in this way up to an energy of about 30,000 wave numbers (cm^{-1}). This procedure provides good free ion wavefunctions, which can then be used to describe the energy levels observed in the solids. These wavefunctions are usually given as a linear combination of Russell–Saunders basis set functions.

1.3 Ions in Crystals

A typical absorption or emission spectrum of an RE ion in a crystal in the visible region has the following features: It shows groups of very sharp lines with an overall group spacing of about 300 cm^{-1}, and the various groups of absorption lines are generally separated by at least 1000 cm^{-1}. Figure 1 shows a wide-scan spectrum of erbium ethylsulfate ($Er(C_2H_5SO_4)_3 \cdot 9H_2O$), which demonstrates exactly the properties mentioned. The transitions are all from the ground term $^4I_{15/2}$ to the excited terms by which the various line groups are designated. The symbols are those of the Russell–Saunders approximation. The inserts at the bottom

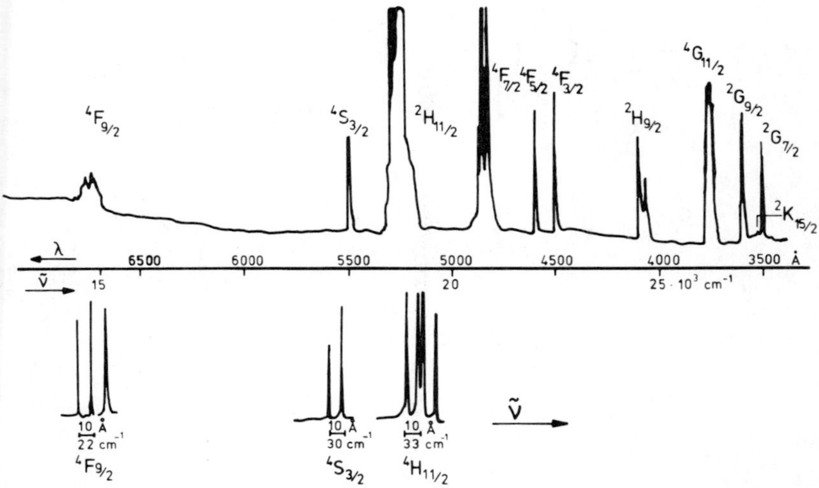

Fig. 1. Typical absorption spectrum of an RE compound: $Er(C_2H_5SO_4)_3 \cdot 9H_2O$. Upper trace: the absorption coefficient as a function of wavelength (wave number) for the transition from the $^4I_{15/2}$ ground term to the excited terms $^4F_{9/2}$, $^4S_{3/2}$, . . . , at 77 K. Insert: the $^4I_{15/2} \rightarrow {}^4F_{9/2}$, $^4S_{3/2}$, and $^2H_{11/2}$ transitions as observed with larger dispersion at T = 4.2 K. The individual lines are about 1 Å wide. The classification of the terms is with Russell–Saunders symbols.

show three transitions, on a much finer energy scale at low temperatures, that clearly exhibit the sharpness of the optical absorption lines.

In order to demonstrate at the outset the resolution possible in RE crystal spectra, we show a group of absorption lines on a still finer scale in Fig. 2, the absorption spectrum of the $^5I_8 \rightarrow {}^5F_5$ transition in holmium ethylsulfate ($Ho(C_2H_5SO_4)_3 \cdot 9H_2O$). The top of Fig. 2 shows the spectrum as taken with a grating spectrograph, exhibiting a number of sharp lines, as in the inserts in Fig. 1. In the lower part of Fig. 2 a Fabry–Pérot interferometer has been placed in front of the grating spectrograph. The resolution in the spectra is thereby increased by about a factor of 10^2. Perpendicular to the spectrum we can see various orders of the Fabry–Pérot interferogram. This shows that a number of lines are split into eight components, which is caused by the interaction of the 4f electrons with the nuclear moment of the Ho nucleus (Ho^{165} nuclear spin $I = \frac{7}{2}$) (see also Section 1.7 and Chapter 7).

The spectra suggest that the various groups of lines are due to transitions between free ion states split by the interaction of the RE ions with the surrounding charges of the lattice. This gives a measure of the magnitude

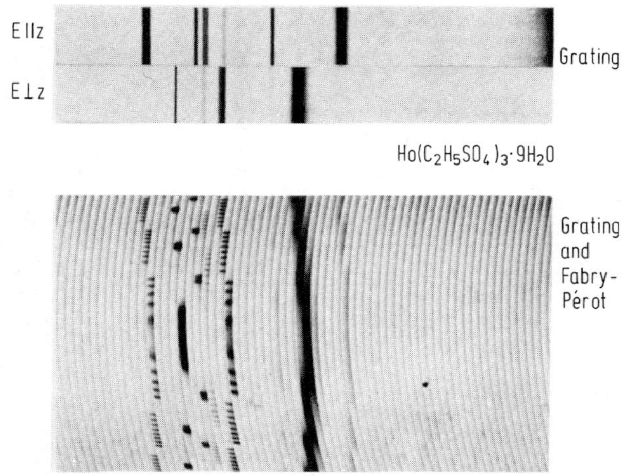

Fig. 2. Absorption spectrum of the $^5I_8 \rightarrow {}^5F_5$ transition in $Ho_{0.05}$: $Y(C_2H_5SO_4)_3 \cdot 9H_2O$ at 4.2 K [HeHü67].

Upper: an absorption spectrum taken with a grating spectrograph; lower: a spectrum where a Fabry–Pérot interferometer has been placed in front of the grating spectrograph (its interference fringes are perpendicular to the grating spectrum, and the different orders are visible). The hyperfine structure due to the large Ho^{165} ($I = 7/2$) nuclear moment is clearly visible as a splitting of each crystal field line into eight hyperfine components (nuclear spin $I = 7/2$ for Ho^{165}). **E** is the electric vector of the radiation, z the hexagonal axis of the crystal.

of the interaction of 4f electrons with the crystal field: it is of the order of the splitting of the line groups, namely, a couple of 100 cm^{-1}. The interaction between the 4f electrons and the crystal is usually described by the field produced by the charges of the crystal at the site of the 4f electrons; henceforth we shall call this field the crystal field.

If these suggestions are correct, they give rise to two questions. The assumed interpretation supposes that the transitions are between 4f states. Such transitions are forbidden for electric dipole radiation by the parity selection rule. What then makes these transitions visible nevertheless? As early as 1937, van Vleck [vVl37] noted that in crystals the interaction with the crystal field or with the lattice vibrations can mix states of different parity into the 4f states, and that these admixtures of other parity wavefunctions make the transitions possible. [There are of course the few cases in which transitions are due to magnetic dipole radiation, which is not forbidden by the parity selection rule; for this radiation the crystal field levels to the lower ones via spontaneous phonon emission. In addition, the linewidth temperature dependence of these crystal field levels is also caused by electron–phonon interaction and can be explained on the basis of the same well-understood processes that determine the spin–lattice interaction in the ground state [MaOr66]. Analysis of the linewidths and their temperature dependence can therefore give added information on the electron–phonon interaction in ionic crystals.

In a number of crystals, very strong emission from certain crystal field levels has been observed, and some of these crystals have therefore been successfully used as lasers. The decay time of these excited levels is almost temperature independent, suggesting that the depopulation of the level takes place by spontaneous emission of radiation. In those cases where a temperature dependence of fluorescence lifetimes has been found, it could be explained by the assumption that in addition to the radiative decay there is also a nonradiative decay by optical phonons. The nonradiative decay occurs by the simultaneous emission of so many phonons that the gap to the next crystal field multiplet can be bridged.

1.5 Energy Transfer

In the preceding section the mechanisms by which an RE system can exchange energy with the lattice were discussed. The possibility also exists, however, that RE ions exchange energy among themselves. Generally speaking, in a solid, by definition, every state is a state of the whole crystal, and thus an absorption or emission of energy involves all the ions of the crystal. That this is usually not the case for RE ions has already been discussed; in some instances, however, the interaction between ions can be large enough so that excitation energy can be transferred from one ion to another. Experimentally this can be analyzed in a crystal containing two or more species of RE ions by exciting one type of ion (donor) with an energetically narrow band of radiation and then observing that the other type of ion (acceptor) emits the radiation. In the simplest form, the energy of the donor and the acceptor is the same (resonant energy transfer). However, energy transfer from the donor to the acceptor can also be observed if the energies do not coincide. The mismatch in energy in such cases is compensated by the phonon system, meaning that a phonon is either absorbed or emitted in order to achieve energy conservation (nonresonant or phonon-assisted energy transfer).

The possibility of transferring energy to an acceptor presents another means of decay of the donor, in addition to the possibility of decay by radiative and nonradiative processes. Therefore, the time dependence of its radiative decay curve is modified, and this modification makes possible a study of the energy transfer processes. In the case of high donor concentrations, the possibility of energy exchange within the donor system also exists. Besides the energy transfer processes, there are those processes in which two ions participate in the emission or absorption of one photon.

1.6 Magnetic Interactions

The size of magnetic interactions in trivalent RE insulators is considerably smaller than that normally observed in similar compounds with 3d transition metal ions. As an example, the Néel temperature of gadolinium oxide (Gd_2O_3) is less than 4 K, whereas that of iron oxide (α-Fe_2O_3) is of the order of 950 K. This is a direct consequence of the localization of the wavefunction of the 4f electrons and of a small hybridization with the orbitals of the ligands. This is, of course, not quite the case for RE semiconductors and metals, where larger exchange interactions can be observed.

Because the linewidth of the 4f transitions is so very small, these magnetic interactions can still be investigated in considerable detail by optical spectroscopy. Often the phenomena observed in that area are described with a molecular field picture. Yet the contribution by anisotropic exchange interaction can be very large, and in some cases application of spin-wave theory has been necessary to explain the spectra completely.

If the magnetic interactions in a compound are of the molecular field type, this gives rise to some very simple spectral features. Basically, spectra obtained under these conditions are identical to those seen in an external magnetic field. This enables us to use the spectra to deduce the magnitude of the molecular field and the temperature dependence of the sublattice magnetization, data that are not otherwise so easily obtainable, especially for antiferromagnetics. A number of the magnetic RE compounds are Ising systems. For this type of interaction the lifetime of the spins in a particular direction can be fairly long, longer than the time given by the inverse of the magnetic splitting energy. Apparently then even above the magnetic ordering temperature, magnetic interactions can be observed that are produced by the local environment of each particular spin. The splitting of the absorption lines observed measures these localized spin configurations rather than the long range magnetization. In these cases, therefore, optical experiments measure a quantity that can also be extracted from diffuse neutron scattering experiments, namely, the short range order parameter.

Sometimes the magnetic interactions in RE compounds take place along chains in the crystal. This means that each magnetic ion interacts predominantly with its two nearest neighbors. In spectra this gives rise to the triplet structure seen in a number of crystals. Again in these cases the spectra enable us to evaluate the interaction constants along the chains of interacting ions.

Of particular interest are the mixed oxides of RE ions and transition metal ions. The most commonly known crystals of this type are perovskites and garnets. Garnets have attracted particular attention because they show very large anisotropic exchange interactions. These anisotropic exchange interactions can be detected directly because they manifest themselves by the fact that the magnetic moment tensor and the magnetic splitting tensor of a particular level no longer coincide.

1.7 Hyperfine Interactions

Because the lines of the REs in crystals are very narrow, it is in principle also possible to detect hyperfine interactions by optical methods (Fig. 2). So far, the very few cases that have been investigated in detail have given useful information on the Sternheimer shielding factors. In addition, in one case [GrHü67] an isotopic shift has also been observed. But so far these investigations cannot compete with the wealth of information obtained from electron spin resonance (ESR) investigations [AbBl70] or Mössbauer effect measurements.

In addition to direct optical spectroscopy, hyperfine interactions in RE salts have also been investigated by an optical microwave double-resonance technique and by various techniques employing narrow line lasers. Especially these latter techniques, which have only recently come into use, will certainly be employed for optical hyperfine structure investigations in the future.

1.8 Jahn–Teller Effect

So far the effect of phonons on the energy of crystal field levels has not been mentioned. The Jahn–Teller theorem [JaTe37], however, requires that for even electron systems with an at least twofold degenerate ground state and for odd electron systems that have a fourfold degenerate ground state, the electron–phonon interaction removes the orbital degeneracies at $T = 0$, thus leading to two singlets and two doublets, respectively. The removal of the degeneracies is accompanied by a lowering of the crystal symmetry; this takes place by a distortion of the lattice and the transition temperature for this crystallographic change is called T_D.

Experience shows that for most crystals the Jahn–Teller splitting is not noticeable; in recent years one class of transparent RE crystals, the RE zircons, has been found that shows a large Jahn–Teller effect. One can visualize the effect in the following way. The curve for the potential energy of an RE ion in a crystal may have two minima, where the barrier between the two minima has a height of the order of kT_D. At temperatures larger than kT_D the thermal vibrations make the system oscillate between the two minima. Then averaged over these oscillations, the ion has a position exactly between the two minima of the potential energy; this symmetric position with respect to the two minima is the degenerate state. At $T < T_D$ the ion becomes stabilized in one of the minima, which corresponds to a lowering of the symmetry at the site of the RE ion. This is a cooperative effect in which all the RE ions of the crystal take part, thus lowering the symmetry of the crystal. This in turn leads to a splitting of the orbital degenerate level, which can be observed by optical means.

If below T_D the ground state has a magnetic moment, the charge distribution of the RE ions can be "pulled" by the application of a magnetic field, from one minimum into the other, as can be seen by the change in direction of the principal axis of the g-factor as a function of the direction of the magnetic field with respect to the crystal axis.

1.9 Divalent Ions

In principle all that has been said so far about trivalent RE ions is also valid for divalent RE ions. Yet the difference in valency causes some changes in the optical spectra. Because of the additional 4f electron in the divalent ions their 4f radius is larger. This causes smaller Coulomb and spin–orbit coupling parameters for the 4f electrons, but also a smaller energy separation between the energy levels of the 4f and 5d electrons. This fact moves transition of the $4f^N \rightarrow 4f^{N-1}5d$ type to lower energies compared to where they are found in the corresponding trivalent compounds. These transitions are not parity forbidden as are the intra-4f transitions and therefore are quite strong; in many instances they can obscure the intra-4f transitions.

Analysis of the intra-4f transitions of divalent RE ions follows the same pattern as that of the trivalent ones, and therefore will not be repeated. Some principle features of the $4f^N \rightarrow 4f^{N-1}5d$ transitions will be discussed with reference to a specific class of compounds.

The Eu chalcogenides constitute an interesting class of materials, many of whose properties depend strongly on the 5d electron. These compounds are magnetic semiconductors, and although their low ordering temperatures make them so far unsuited for technical applications, they nevertheless serve to elucidate the properties of magnetic semiconductors. In the present context the most important phenomenon is the occurrence of a large red shift of the absorption edge in going through the magnetic transition temperature.

1.10 Rare Earths in Glasses

In this book we concentrate on solids in their crystalline form. Stimulated by the successful operation of RE glass lasers, researchers have investigated the optical properties of REs in glasses. In these hosts the local environment of an RE ion is roughly the same as that in a crystal, giving rise to a spectrum consisting of distinct lines. Since the far field is, however, different for every RE ion, the lines show a large inhomogeneous broadening, typically of the order of 100 cm^{-1}. This characteristic makes the spectra much less useful for an investigation of "crystal field" properties. Therefore most investigations of the optical properties of RE-doped glasses deal with the determination of lifetimes and oscillator strength, which are similar to those observed in crystals.

1.11 Rare Earth Lasers

Soon after the successful operation of the ruby laser, lasers of crystals doped with RE ions were also constructed. We shall therefore briefly discuss the principles of laser operation and then list some of the lasers using REs. By far the most important RE laser material is Nd-doped glass, which is used, for example, in controlled nuclear fusion experiments. The glass host has several advantages over conventional crystalline materials; it is easy to dope homogeneously, easy to handle, and can be produced in quite large volumes. Also, Nd glass lasers have been used in optoelectronic devices in the hope that these lasers can be pumped by light-emitting diodes.

CHAPTER 2

Free Ions

2.1 Introduction

The subject of this chapter has been dealt with extensively by Wybourne [Wy65]. We refer the reader to his book for details, especially with respect to theoretical questions. Here we want to sketch only the main facts that are important for an understanding of the energy structure of RE ions in solids.

The REs in solids are either divalent or trivalent. Their electronic configuration is $4f^N5s^25p^6$ or $4f^{N-1}5s^25p^6$, respectively. By far the most common valence state of the RE ions in solids is the trivalent one. When the term RE is used henceforth, we shall always mean the trivalent state unless otherwise specified. The much less common divalent state will be dealt with in Chapter 9.

The 4f electrons are not the outermost ones. They are "shielded" from external fields by two electronic shells with larger radial extension ($5s^25p^6$), which explains the "atomic" nature of their spectra. Thus the 4f electrons are only weakly perturbed by the charges of the surrounding ligands. From another point of view, an electronic shell in a solid is the more atomiclike the deeper its position below the top of the valence band, which is the highest occupied level. Photoemission experiments [WeCo71] have shown that in trivalent RE compounds the position of the 4f states is at least 5 eV below the top of the valence band. Therefore, these states will be quite atomiclike. This is why the RE ions are such a useful probe in a solid: The crystal environment constitutes only a small perturbation on the atomic energy levels, and many of the solid state, and hence spectroscopic, properties can be understood from a consideration of the free ions. In turn, the wavefunctions of the free ions constitute a good zero order approximation for a description of solid state properties.

The energy levels of the 4f configuration for trivalent free RE ions have been analyzed in arc spectra for Pr^{3+} [Su65], Gd^{3+} [KiCr70], and Er^{3+}[CrMo67]. Tables 1 and 2 list the experimentally observed energy levels together with the centers of gravity for $PrCl_3$ and Er^{3+} in a number of host materials. The Russell–Saunders notation for the energy levels (terms) is used; the exact meaning of these symbols will be explained later. The remarkable fact immediately apparent from these two tables is that the positions of the terms in the free ion and in various ionic hosts are the same to within a few hundred wave numbers. In looking over the numbers we find that in most cases agreement is within 100 cm^{-1}, and only rarely do the differences amount to 200 cm^{-1} or more. This finding had been anticipated from the electronic structure but is nicely confirmed by the experimental data. To emphasize this, Fig. 3 shows for Er^{3+} the position of the free ion terms and those obtained from the optical spectra of $Er(C_2H_5SO_4)_3 \cdot 9H_2O$ (solid). This figure also shows theoretical results for the energy levels. It can be seen that an "advanced" theory (intermediate coupling) describes the energies very well; the Russell–Saunders approximation yields less satisfactory results.

TABLE 1

ENERGY LEVELS FOR Pr^{3+} AS OBTAINED FROM ARC SPECTRA YIELDING THOSE OF THE Pr^{3+} ION, AND FOR $PrCl_3$[a]

	E (Vapor state) (cm^{-1})	E (Crystal) (cm^{-1})		E (Vapor state) (cm^{-1})	E (Crystal) (cm^{-1})
$^3H_4$	0	0	$^1G_4$	9921.4	9697.6
$^3H_5$	2152.2	2117.4	$^1D_2$	17334.5	16639.3
$^3H_6$	4389.1	4306.3	$^3P_0$	21390.1	20383.4
$^3F_2$	4996.7	4846.6	$^3P_1$	22007.6	20984.9
$^3F_3$	6415.4	6232.3	$^1I_6$	22211.6	21324.5
$^3F_4$	6854.9	6681.7	$^3P_2$	23160.9	22139.1
			$^1S_0$		47200.0[b]

[a] From [Su65].

[b] From [Lo67] measured in CaF_2.

TABLE 2

Energy Levels for Er^{3+} as Obtained from Arc Spectra, and Energy Levels in Crystalline Hosts[a]

Term	Free ion	LaF_3	ErES[b]	$ErCl_3 \cdot 6H_2O$	$LaCl_3$	$LaBr_3$	Y_2O_3
$^4I_{15/2}$	0.0	0.0	0.0	0.0	0.0	0.0	0.0
$^4I_{13/2}$	6485	6480			6481	6475	6458
$^4I_{11/2}$	10,123	10,123	10,113	10,109	10,111		10,073
$^4I_{9/2}$	12,345	12,350	12,366	12,349	12,351	12,338	12,287
$^4F_{9/2}$	15,182	15,235	15,207	15,182	15,175	15,149	15,071
$^4S_{3/2}$	18,299	18,353	18,327	18,284	18,290	18.260	18,072
$^2H_{11/2}$	19,010		19,087	19,055			18,931
$^4F_{7/2}$	20,494	20,492	20,457	20,426			20,267
$^4F_{5/2}$	22,181	20,161	22,121	22,078	22,067	22,021	21,894
$^4F_{3/2}$	22,453	22,494	22,461	22,436	22,409	24,369	22,207
$^2H_{9/2}$	24,475	24,526	24,515	24,464	24,433		24,304
$^4G_{11/2}$	26,376	26,368	26,348	26,297	26,271	26,180	26,074
$^2G_{9/2}$	27,319	27,412	27,360	27,285		27,159	
$^2K_{15/2}$	27,584		27,660	27,649			
$^2G_{7/2}$	27,825	28,081	27,970	27,940			
$^2P_{3/2}$	31,414	31,501	31,480		31,384	31,284	31,186
$^2P_{1/2}$				32,630			
$^2K_{13/2}$			32,960				
$^4G_{5/2}$			33,250				
$^4G_{7/2}$	33,849	33,995	33,930				33,697
$^2D_{5/2}$			34,810				
$^4G_{9/2}$			36,370				

[a] From [CrMo67, Hü62a].
[b] $Er(C_2H_5SO_4)_3 \cdot 9H_2O$.

The energy levels of a free RE ion are usually interpreted by considering only interactions between the 4f electrons themselves. Since all the other electronic shells are spherically symmetric, their effect on all the terms of a 4f configuration is the same in first order. The situation changes if we take into account that the aspherical 4f shell (because it is nonclosed) can distort the closed spherical shells. Then they can also interact with the 4f shell. This is, however, a small effect and will be neglected in our treatment.

We remind the reader that the interpretation of experimental data in terms of a quantum mechanical calculation can usually be separated into two parts. One consists in solving the angular part of the Schrödinger equation, and if the symmetry of the problem is known, this can often be done exactly, albeit by very cumbersome procedures. This step provides information on the number of energy levels expected for a certain problem

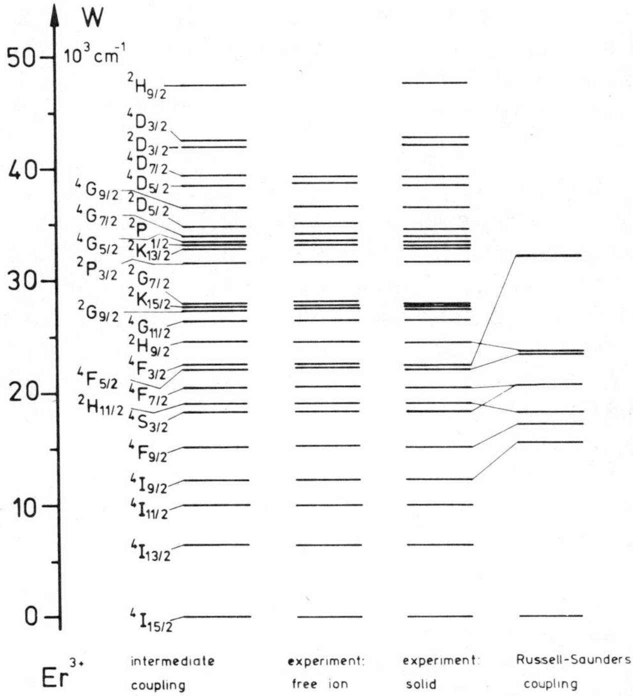

Fig. 3. Comparison of energy levels of Er^{3+} from theory (intermediate coupling); arc spectra (free ion); crystal spectra ($Er(C_2H_5SO_4)_3 \cdot 9H_2O$; approximate theory (Russell–Saunders coupling; for instructive purposes only a few examples are shown).

and on their relative ordering, as well as on the selection rules for transitions of various types of radiation. The second part involves calculating the radial part of the wavefunction, which requires knowledge of the exact magnitude of the wavefunctions and is much harder to do. To check the internal consistency of an interpretation, especially when dealing with atomic energy levels, it is customary to use the radial integrals as adjustable parameters and solve the angular part of the wavefunction exactly. We shall see that this procedure is also adopted in the field we are dealing with.

Having recognized that all the electronic shells except the 4f shell are spherically symmetric, and therefore do not contribute significantly to the relative positions of the 4f energy levels, we can write the Hamiltonian that determines the 4f energy levels as

$$\mathscr{H} = -\frac{\hbar^2}{2m}\sum_{i=1}^{N}\Delta_i - \sum_{i=1}^{N}\frac{Z^*e^2}{r_i} + \sum_{i<j}^{N}\frac{e^2}{r_{ij}} + \sum_{i=1}^{N}\zeta(r_i)\mathbf{s}_i \cdot \mathbf{l}_i \qquad (2.1)$$

where $N = 1, \ldots, 14$ is the number of the 4f electrons, Z^*e the screened charge of the nucleus because we have neglected the closed electronic shells, and $\zeta(r_i)$ the spin–orbit coupling function

$$\zeta(r_i) = \frac{\hbar^2}{2m^2c^2r_i}\frac{dU(r_i)}{dr_i}$$

where $U(r_i)$ is the potential in which the electron i is moving. The first term in the Hamiltonian (2.1) is the kinetic energy of the 4f electrons and the second term represents their Coulomb interaction with the nucleus. Because there is an interaction with the closed shells that modifies only this term's magnitude but not its symmetry, we replace the real charge of the nucleus by a screened charge. These first two terms of the Hamiltonian are spherically symmetric and therefore do not remove any of the degeneracies within the configuration of the 4f electrons. We can neglect them in the following discussion.

The next two terms, which represent the mutual Coulomb interaction of the 4f electrons (H_c) and their spin–orbit interaction (H_{so}), are responsible for the energy level structure of the 4f electrons. In atomic theory there are two limiting cases for the relative size of these two interactions; for $H_c \gg H_{so}$ we have the so-called Russell–Saunders coupling, where the spin–orbit interaction is only a small perturbation on the energy level structure which has been determined from the diagonalization of H_c. This is true for light elements and for a long time was also assumed to hold for the REs. In the second limiting case the Coulomb interaction is small compared to the spin–orbit interaction and we have the so-called *j–j* coupling scheme. Both limiting cases are relatively easy to deal with theoretically and can be solved by perturbation theory. In REs, unfortunately, the two last terms in Eq. (2.1) are of about equal magnitude and the energy level calculations are therefore mathematically more involved; this situation is called intermediate coupling.

The Hamiltonian in eq. (2.1) is solved in a one-electron approximation; that is, electron correlations are neglected. To calculate the energy levels in the intermediate coupling approximation we have to calculate the matrix elements for the Hamiltonian

$$H_1 = H_c + H_{so}$$

in a set of basis functions and then diagonalize the matrix for the specific $4f^N$ configuration that interests us. It is now common practice to use a basis set of Russell–Saunders eigenfunctions. The Hamiltonian H_1 is diagonal in J, and therefore the total matrix of energies for the $4f^N$ configuration can be split up into submatrices for states with the same J. These states (terms) are then still degenerate in M_J and are a linear combination of states (terms) with different L and S but the same J. There are also configurations with the same L and S that occur more than once; to distinguish them, new quantum numbers must be introduced. (For details see the book by Judd [Ju63]). Details of the matrix element calculations of H_c and H_{so} will not be given here, since there are many excellent reviews in which the derivations can be found [Ju63, Wy65].

Some of the formulas needed for the calculations are presented because they serve to introduce the concept of tensor operators, a technique that has many applications. Therefore, the emphasis will be on showing how the results of this mathematical tool can be used rather than on a rigorous derivation of the formulas, which can be found in a number of books [Ed60, Ro57, FaRa59]. The basic theory for a description of atomic spectra is given by Condon and Shortley [CoSh35], and the tensor operator algebra was developed by Racah [Ra42, Ra43, Ra49].

Let us consider as an example the configuration of two f electrons. The Pauli exclusion principle shows that in the Russell–Saunders coupling scheme the states ^1I, ^3H, ^1G, ^3F, ^1D, ^3F, and ^1S are possible. Hund's rule requires that $^3H_4$ be the ground term; but in order to obtained the energy position of the other terms, we have to calculate the Coulomb and spin–orbit matrix elements for the f^2 configuration.

The Coulomb interaction matrix elements have the form

$$(f^2SLM_SM_L|(e^2/r_{12})|f^2S'L'M'_SM'_L) \tag{2.2}$$

We can develop $1/r_{12}$ in Legendre polynomials, obtaining

$$\frac{1}{r_{12}} = \sum_k \left(\frac{r_<^k}{r_>^{k+1}}\right) P_k(\cos \omega_{12}) \tag{2.3a}$$

where $r_<$ and $r_>$ are the smaller and larger of the radii of the two electrons, and

$$P_k(\cos \omega_{12}) = \frac{4\pi}{2k+1} \sum_q Y_{kq}^*(\vartheta_1, \varphi_1) Y_{kq}(\vartheta_2, \varphi_2) \tag{2.3b}$$

where Y_{kq} are spherical harmonics (for a definition see, e.g., [Ed60]). Now define the operators

$$C_{kq} = \left(\frac{4\pi}{2k+1}\right)^{1/2} Y_{kq} \tag{2.4}$$

(C_{kq} is called a tensor operator),† with the result

$$P_k(\cos \omega_{12}) = \sum_q (-1)^q (C_{k-q}(1) \cdot C_{kq}(2)) = \mathbf{C}_k(1) \cdot \mathbf{C}_k(2) \tag{2.5}$$

and the Coulomb matrix element is

$$\sum_k (\mathrm{f}^2 SLM_S M_L | (e^2 r_<^k / r_<^{k+1}) \mathbf{C}_k(1) \cdot \mathbf{C}_k(2) | \mathrm{f}^2 S'L'M_S' M_L') \tag{2.6}$$

Evaluation of the Coulomb matrix element involves two steps: First, we calculate the radial part, which we shall deal with later; then we evaluate the angular part, which is our main concern here. The calculation of matrix elements of tensor operators like the C_{kq} (or of their products) is a problem encountered quite often in physics. Most of the numerical work connected with these calculations can be avoided these days because of the tables of Rotenberg *et al.* [RoBi59] and Nielson and Koster [NiKo64]. To make this book reasonably self-contained, we shall present some of the principal formulas for tensor operator algebra.

2.2 Matrix Elements of Tensor Operators

The foundations for the formulas given in this section are found in [Ra42, Ra43, Ro57, Ed60, FaRa59, Ju63]. Our approach is that of Edmonds and Judd; see their book for details.

We want to compute the matrix elements of tensor operators in momentum space. We start with the equation

$$C_{kq} = \left(\frac{4\pi}{2k+1}\right)^{1/2} Y_{kq} \tag{2.4}$$

for a tensor operator.

To find the commutation relations of this operator with J_x, J_y, J_z, we have only to find those for the angular momentum operator $\mathbf{l}$ because this operator is the only part of $\mathbf{J}$ that does not commute with C_{kq}; with

$$l_z = \frac{1}{i}\frac{\partial}{\partial\varphi} \quad \text{and} \quad l_\pm = e^{\pm i\varphi}\left(\pm\frac{\partial}{\partial\vartheta} + i\cot\frac{\partial}{\partial\varphi}\right) \tag{2.7}$$

† We have chosen a different form of indexing than that found in most of the literature, writing C_{kq} instead of $C_q^{(k)}$ in order to avoid any misreading of exponents.

we find

$$[J_z, C_{kq}] = qC_{kq}$$
$$[J_\pm, C_{kq}] = [k(k+1) - q(q \pm 1)]^{1/2} C_{kq\pm 1} \tag{2.8a}$$

(where we assume from now on that the angular momentum is measured in units of $\hbar$).

A general tensor operator T_{kq} or U_{kq} is now defined by its commutation relations, which are the same as those just given for C_{kq}.

$$[J_z, T_{kq}] = qT_{kq}$$
$$[J_\pm, T_{kq}] = [k(k+1) - q(q \pm 1)]^{1/2} T_{kq\pm 1} \tag{2.8b}$$

In the simplest case we have matrix elements, between two states, with momentum j and its projection m. Then the dependence on the projection quantum number m can be calculated with the Wigner–Eckart theorem:

$$(\gamma jm|T_{kq}|\gamma' j' m') = (-1)^{j-m}(\gamma j\|T_k\|\gamma' j')\begin{pmatrix} j & k & j' \\ -m & q & m' \end{pmatrix} \tag{2.9}$$

where $(\| \quad \|)$ is a so-called reduced matrix element and the last factor is a $3j$ symbol. (For a definition see, e.g., [Ed60, Ju63].) γ stands for all quantum numbers on which the tensor operator does not act. The problem now is to calculate the reduced matrix element; to do so, we have to know the nature of T_{kq} and of j. It is customary to normalize the $(l\|T_k\|l)$ in such a way that

$$(l\|T_k\|l') = 1 \tag{2.10a}$$

where l is the angular momentum of an electron. Other useful normalizations are

$$(l\|l\|l') = \delta_{ll'}[l(l+1)(2l+1)]^{1/2}$$
$$(s\|s\|s') = \delta_{ss'}[s(s+1)(2s+1)]^{1/2} \tag{2.10b}$$
$$(l\|C_k\|l') = (-1)^l[(2l+1)(2l'+1)]^{1/2}\begin{pmatrix} l & k & l' \\ 0 & 0 & 0 \end{pmatrix}$$

As we saw in Section 2.1, the states of the free RE ion system are classified by the quantum numbers S, L, and J. It is often necessary to calculate the matrix elements of an operator that acts on only a specific part of a system. For example, the operator of the crystal field acts on only the orbital angular momentum and is diagonal in the spin. (Conversely, the exchange interaction operator acts only on the spin and not on the orbit.) The most general instance of this problem is the case where we have a product of tensor operators, one of which acts only on one part of the system and the other only on the rest. The formulas for the matrix elements in these cases are given here. The product tensor operator X_{KQ} is defined by

$$\begin{aligned} X_{KQ} &= \sum_{q_1, q_2} T_{k_1 q_1} U_{k_2 q_2} (2K+1)^{1/2} (-1)^{k_1 - k_2 + Q} \begin{pmatrix} k_1 & k_2 & K \\ q_1 & q_2 & -Q \end{pmatrix} \\ &= \{\mathbf{T}_{k_1} \cdot \mathbf{U}_{k_2}\}_{KQ} \end{aligned} \tag{2.11}$$

where $T_{k_1 q_1}$ acts only on part 1 and $U_{k_2 q_2}$ only on part 2. X_{KQ} obeys the commutation relations

$$\begin{aligned} [J_z, X_{KQ}] &= Q X_{KQ} \\ [J_\pm, X_{KQ}] &= [K(K+1) - Q(Q \pm 1)]^{1/2} X_{KQ\pm 1} \end{aligned} \tag{2.12}$$

If $k_1 = k_2 = k$, it is sometimes useful to define the scalar product of two tensor operators by the equation

$$(\mathbf{T}_k \cdot \mathbf{U}_k) = \sum_q (-1)^q T_{kq} U_{k-q} \tag{2.13a}$$

or

$$(\mathbf{T}_k \cdot \mathbf{U}_k) = (-1)^k (2k+1)^{1/2} \{\mathbf{T}_k \cdot \mathbf{U}_k\}_{00} \tag{2.13b}$$

For the calculation of the matrix elements of X_{KQ} the part that contains the projection quantum numbers Q can again be obtained from the Wigner–Eckart theorem (2.9):

$$\begin{aligned} &(\gamma j_1 j_2 J M_J | X_{KQ} | \gamma' j_1' j_2' J' M_J') \\ &\quad = (-1)^{J - M_J} \begin{pmatrix} J & K & J' \\ -M_J & Q & M_{J'} \end{pmatrix} (\gamma j_1 j_2 J \| X_K \| \gamma' j_1' j_2' J') \end{aligned} \tag{2.14}$$

(j_i and j_2 the momenta of the two systems couple to J the momentum of the total system.)

The reduced matrix element can then be expressed in products of the reduced matrix elements of T_{k_1} and U_{k_2} (T_{k_1} acts only on part 1 and U_{k_2} acts only on part 2 of the system)

$$\begin{aligned} &(\gamma j_1 j_2 J \| X_K \| \gamma' j_1' j_2' J') \\ &\quad = \sum_{\gamma''} (\gamma j_1 \| T_{k_1} \| \gamma'' j_1')(\gamma'' j_2 \| U_{k_2} \| \gamma' j_2') \\ &\qquad \times \{(2J+1)(2K+1)(2J'+1)\}^{1/2} \begin{Bmatrix} j_1 & j_1' & k_1 \\ j_2 & j_2' & k_2 \\ J & J' & K \end{Bmatrix} \end{aligned} \tag{2.15}$$

where the last factor is a 9j symbol the definition of which is given by Edmonds [Ed60]. For the case in which $k_1 = k_2 = k$, $K = 0$, and using the definition for the scalar product of two tensor operators, eq. (2.15) gets much simpler (especially because the resulting matrix elements are diagonal in J and M_J):

$$(\gamma j_1 j_2 J M_J|(\mathrm{T}_k \cdot \mathrm{U}_k)|\gamma' j_1' j_2' J' M_J')$$
$$= (-1)^{j_1' + j_2 + J}\delta_{JJ'}\delta_{M_J M_J'}$$
$$\times \begin{Bmatrix} j_1' & j_2' & J \\ j_2 & j_1 & k \end{Bmatrix} \sum_{\gamma''} (\gamma j_1 \| T_k \| \gamma'' j_1')(\gamma'' j_2 \| U_k \| \gamma' j_2') \qquad (2.16)$$

where { } is a 6j symbol [Ed60].

From eq. (2.15) we obtain the useful equation for a tensor operator T_k acting only on part 1 in the coupled system $(\gamma j_1 j_2 J M_J)$ and for the operator U_k acting only on part 2 in the same system; setting $k_2 = 0$ $(k_1 = 0)$, we obtain

$$(\gamma j_1 j_2 J \| T_k \| \gamma' j_1' j_2' J') = \delta_{j_2 j_2'}(-1)^{j_1 + j_2 + J' + k}\{(2J + 1)(2J' + 1)\}^{1/2}$$
$$\times \begin{Bmatrix} J & k & J' \\ j_1' & j_2 & j_1 \end{Bmatrix} (\gamma j_1 \| T_k \| \gamma' j_1') \qquad (2.17a)$$

and

$$(\gamma j_1 j_2 J \| U_k \| \gamma' j_1' j_2' J') = \delta_{j_1 j_1'}(-1)^{j_1' + j_2 + J + k}\{(2J + 1)(2J' + 1)\}^{1/2}$$
$$\times \begin{Bmatrix} J & k & J' \\ j_2' & j_1 & j_2 \end{Bmatrix} (\gamma j_2 \| U_k \| \gamma' j_2') \qquad (2.17b)$$

All these equations are presented without proof. They are mainly intended to provide the newcomer to the field with a feeling for the procedures involved and to enable him to apply the relevant tables [RoBi59] and [NiKo64]. By no means can these few formulas take the place of a thorough study of the literature for those who want to perform original calculations.

2.3 Coulomb and Spin–Orbit Energies; Intermediate Coupling

With the formulas in the preceding section, we are in a position to obtain the matrix elements of the Coulomb and spin–orbit interactions. We shall give brief examples for the f^2 configuration.

The Coulomb interaction is an electric interaction and therefore does not act on the spin; thus we obtain [using eq. (2.16)]

$$(\mathrm{f}^2SLM_SM_L|e^2/r_{12}|\mathrm{f}^2SL'M_SM'_L)$$

$$= \sum_k \mathrm{f}_k(r)(\mathrm{f}^2\cdots|\mathbf{C}_k(1)\cdot\mathbf{C}_k(2)|\mathrm{f}^2\cdots)$$

$$= \sum_k \mathrm{f}_k(r)(-1)^{2l+L}\delta_{LL'}\delta_{M_LM'_L}\delta_{SS'}\delta_{M_SM'_S}\begin{Bmatrix} l & l & k \\ l & l & L \end{Bmatrix}(l\|C_k\|l)^2$$

$$= \sum_k \mathrm{f}_k(-1)^L\delta_{LL'}\delta_{M_LM'_L}\delta_{SS'}\delta_{M_SM'_S}7^2\begin{pmatrix} 3 & k & 3 \\ 0 & 0 & 0 \end{pmatrix}^2\begin{Bmatrix} 3 & 3 & k \\ 3 & 3 & L \end{Bmatrix} \quad (2.18)$$

where for the radial integral we used the common definition

$$\mathrm{f}_k(r) = F^k = e^2\int(r_<^k/r_>^{k+1})R_{4\mathrm{f}}^2(r_1)R_{4\mathrm{f}}^2(r_2)\,dr_1\,dr_2 \quad (2.19)$$

Applying the formulas to the states of the f^2 configuration and using the tables of Rotenberg *et al.* [RoBi59], we obtain

$$W_c(^3\mathrm{H}) = F_0 - 25F_2 - 51F_4 - 13F_6$$

$$W_c(^3\mathrm{F}) = F_0 - 10F_2 - 33F_4 - 286F_6$$

$$W_c(^1\mathrm{G}) = F_0 - 30F_2 + 97F_4 + 78F_6$$

$$W_c(^1\mathrm{D}) = F_0 + 19F_2 - 99F_4 + 715F_6$$

$$W_c(^3\mathrm{P}) = F_0 + 45F_2 + 33F_4 - 1287F_6$$

$$W_c(^1\mathrm{I}) = F_0 + 25F_2 + 9F_4 + F_6$$

$$W_c(^1\mathrm{S}) = F_0 + 60F_2 + 198F_4 + 1716F_6$$

where in order to avoid large denominators new parameters have been introduced:

$$F^0 = F_0, \quad F^2 = 225F_2, \quad F^4 = 1089F_4, \quad F^6 = (18{,}4041/25)F_6 \quad (2.20)$$

In principle we can calculate the Coulomb matrix element for every f^N configuration by tensor operator techniques. For $N > 2$, classification of the states, however, becomes a problem. We need not treat this subject in detail here because the matrix elements of all the states for p^N, d^N, and f^N configurations have been calculated by Nielson and Koster [NiKo64] and are available in table form. There are, however, some conventions that enable us to apply these tables [NiKo64] directly. Instead of writing the Coulomb matrix elements in terms of the parameters F_k (or F^k), Racah [Ra49] found it more convenient to define new parameters and write

$$E = \sum_{k=0}^{3} e_k E^k \tag{2.21a}$$

with

$$\begin{aligned} E^0 &= F_0 - 10F_2 - 33F_4 - 286F_6, \\ E^1 &= \tfrac{1}{9}(70F_2 + 231F_4 + 2002F_6) \\ E^2 &= \tfrac{1}{9}(F_2 - 3F_4 + 7F_6), \\ E^3 &= \tfrac{1}{3}(5F_2 + 6F_4 - 91F_6) \end{aligned} \tag{2.21b}$$

where the numbers in the tables [NiKo64] are given in terms of the E^k.

Conversion from the E^k to F_k parameters is done by the formulas

$$\begin{aligned} F_0 &= \tfrac{1}{7}(7E^0 + 9E^1), \\ F_2 &= \tfrac{1}{42}(E^1 + 143E^2 + 11E^3) \\ F_4 &= \tfrac{1}{77}(E^1 - 130E^2 + 4E^3), \\ F_6 &= \tfrac{1}{462}(E^1 + 35E^2 - 7E^3) \end{aligned} \tag{2.21c}$$

The matrix elements of the spin–orbit interaction can also be evaluated by using the tensor formalism. Again the calculations of Rotenberg *et al.* [RoBi59] and Nielson and Koster [NiKo64] can be used.

The spin–orbit coupling Hamiltonian $\sum_{i=1}^{N} \zeta(r_i)(\mathbf{s}_i \cdot \mathbf{l}_i)$ is a simple example of a product tensor operator; the tensors are of rank one and one acts only on the spin, the other only on the orbit. For this case the formulas simplify considerably and we obtain (see, e.g., [Ju63]), using formulas (2.16) and (2.10b)

$$\begin{aligned} &\left(\mathrm{f}^N SLJM_J \middle| \sum_{i=1}^{N} \zeta(r_i)(\mathbf{s}_i \cdot \mathbf{l}_i) \middle| \mathrm{f}^N S'L'J'M'_J\right) \\ &\quad = \zeta_{4\mathrm{f}}(-1)^{J+L+S'} \delta_{JJ'}\,\delta_{M_JM'_J}\sqrt{84} \begin{Bmatrix} S & S' & 1 \\ L' & L & J \end{Bmatrix} (\mathrm{f}^N SL \| V_{11} \| \mathrm{f}^N S'L') \end{aligned} \tag{2.22}$$

Here the product tensor operator has been defined as

$$V_{1x} = \sum_i (su_x)_i$$

Since u_x is a normalized tensor operator, we have to multiply by the corresponding normalization matrix element (2.10b), whereas the normalization for S is incorporated in the evaluation of V_{1x}; in our case, $V_{1x} = V_{11}$. The matrix elements for the operator V_{11} are calculated in [NiKo64] and thus the spin–orbit matrix elements are obtainable.

As for the spin–orbit coupling parameter, from the assumed equivalence of the 4f electrons it follows that this parameter is also equal for all 4f electrons, and is defined as the radial integral

$$\zeta_{4f} = \int R_{4f}^2 \zeta(r)\, dr \tag{2.23}$$

The matrix elements of the spin–orbit coupling are responsible for the mixing of states with the same J but different L and S. The diagonal matrix elements are obtained by the familiar formula

$$(\mathrm{f}^N \cdots |H_{so}| \mathrm{f}^N \cdots) = \zeta(\lambda/2)[J(J+1) - L(L+1) - S(S+1)] \tag{2.24}$$

where λ is a constant for a state with given L and S. The energy levels calculated from this formula lead to the ordering with the Landé interval rule, namely, that in any Russell–Saunders multiplet the interval between neighboring terms is proportional to the higher J of the pair.

Rare earth ions are characterized by the fact that the Coulomb interaction and the spin–orbit interaction are of the same order of magnitude. This can be visualized in the following way. Ce^{3+} and Yb^{3+} have one electron and one hole in the 4f shell, respectively. Therefore the splitting between the $^2F_{7/2}$ and $^2F_{5/2}$ terms (2000 cm^{-1} for Ce^{3+} and 10,000 cm^{-1} for Yb^{3+}) are produced only by the spin–orbit interaction. On the other hand, the energy separation between, e.g., the $^1D_2$ term and the center of gravity of the ^3P free ion multiplet in Pr^{3+} is 4000 cm^{-1}; this is produced by the Coulomb interaction and the comparison of the mentioned numbers shows the equality in the size of the Coulomb and the spin–orbit energies.

In an actual calculation of the energy levels for a given $4f^N$ configuration we therefore have to set up the matrix for each J with all the Coulomb and spin–orbit matrix elements. The parameters F_2, F_4, F_6, and ζ_{4f} are treated as free parameters that are adjusted to the experimentally observed energy levels. Good starting points for these parameters are the values obtained by using the hydrogenic wavefunction for the 4f electrons. This approach is particularly apt for the ratios F_4/F_2 and F_6/F_2, which are relatively insensitive to the choice of wavefunction. It is therefore customary to express the energies in terms of the Slater integral F_2. We again give some values for f^2. Using formula (2.22) for $N = 2$, we can calculate the spin–orbit matrix elements for the $J = 4$ manifold of f^2, obtaining the matrix [Ju63]

$$\begin{pmatrix} W_c(^3H) - 3\zeta & -(\frac{10}{3})^{1/2}\zeta & 0 \\ -(\frac{10}{3})^{1/2}\zeta & W_c(^1G) & (\frac{11}{3})^{1/2}\zeta \\ 0 & (\frac{11}{3})^{1/2}\zeta & W_c(^3F) + \frac{3}{2}\zeta \end{pmatrix}$$

where the Coulomb matrix elements can be calculated by using, e.g., the hydrogenic F_4/F_2 and F_6/F_2 ratios which are $F_4/F_2 = 0.138$ and $F_6/F_2 = 0.0151$ [Ju63]. Setting $F_2 = 310\ \text{cm}^{-1}$, as for Pr^{3+} [Ju63], we get

$$W_c(^3H) = -32.3F_2 = -9982\ \text{cm}^{-1}$$

$$W_c(^3F) = -18.9F_2 = -5859\ \text{cm}^{-1}$$

$$W_c(^1G) = -15.4F_2 = -4774\ \text{cm}^{-1}$$

For Pr^{3+} the spin–orbit coupling constant is $\zeta = 750\ \text{cm}^{-1}$ [AbBl70]. Inserting the Coulomb energies and ζ into the matrix, we see that the off-diagonal matrix elements are not negligible with respect to the diagonal ones—which makes full diagonalization necessary for the calculation of the energies.

The ground state manifold for Pr^{3+} can be calculated approximately by using eq. (2.24). Realizing that $\lambda = \frac{1}{2}$, using the spin–orbit coupling constant, and setting the energy of the lowest term to zero, we find

$$W(^3H_4) = 0\ \text{cm}^{-1}\ (0\ \text{cm}^{-1}), \qquad W(^3H_5) = 1875\ \text{cm}^{-1}\ (2152\ \text{cm}^{-1})$$

$$W(^3H_6) = 4125\ \text{cm}^{-1}\ (4389\ \text{cm}^{-1})$$

where the number in parentheses is the experimental value. As found frequently for Pr^{3+} too, the ground state multiplet is not disturbed much by the intermediate coupling. The intermediate coupling calculation reduces of course the differences between the theoretical and experimental energies yielding

$$W(^3H_4) = 0\ \text{cm}^{-1}, \qquad W(^3H_5) = 2127\ \text{cm}^{-1},$$

and

$$W(^3H_6) = 4355\ \text{cm}^{-1}$$

A scheme that illustrates well the deviations from pure Russell–Saunders coupling is one in which the energies of the terms obtained by diagonalizing the Coulomb and the spin–orbit interaction matrices of a $4f^N$ configuration for every value of J are plotted as a function of ζ (or rather ζ/F_2). An example is Fig. 4, where these results are shown for the $4f^3$ and $4f^{11} = 4f^{-3}$ configurations representing Nd^{3+} and Er^{3+} [Di68]. The electrostatic matrix elements are the same for a configuration of N electrons and N holes because a two-particle interaction is involved. On the other hand, the spin–orbit interaction changes sign, being negative for the $4f^{-N}$ configuration. The energies of the $4f^N$ and $4f^{-N}$ configurations are thus obtained from the same matrices where only the sign of the spin–orbit matrix elements changes in going from $4f^N$ to $4f^{-N}$. The same plot of the energies as a function of ζ/F_2 therefore contains for $(\zeta/F_2) < 0$, the energies for the $4f^{-N}$ configuration, and for $(\zeta/F_2) > 0$, the energies for the $4f^N$ configuration.

Figure 4 has several interesting features. For small values of ζ/F_2 the energies are approximately a linear function of this parameter, showing negligible mixing of the wavefunctions of different terms, as expected in the Russell–Saunders approximation. For large values of ζ/F_2 strong nonlinearities are observed, showing considerable mixing of the wavefunctions of the terms belonging to the same J manifold. The figure also indicates the values of ζ/F_2 at which the best agreement with experimentally measured energy levels is obtained. For Er^{3+} these energies are also shown in Fig. 3, and good agreement with the experimental data is evident. This figure also contains the energy levels calculated in the pure Russell–Saunders approximation and we note the much worse correspondence with the experimental data.

The wavefunctions are now no longer pure Russell–Saunders wavefunctions, but a linear combination of them. For example, the ground term wavefunction, of Er^{3+} is now in the intermediate coupling approximation [Ka61]

$$|{}^4\mathrm{I}'_{15/2}\rangle = 0.984|{}^4\mathrm{I}_{15/2}\rangle + 0.176|{}^2\mathrm{K}_{15/2}\rangle + 0.019|{}^2\mathrm{L}_{15/2}\rangle$$

(instead of pure ${}^4\mathrm{I}_{15/2}$) where the primed symbol denotes the intermediate coupling state. In this case the admixture is small and can almost be neglected. This is not so for some of the higher terms for example [Ka61]

$$\begin{aligned}|{}^4\mathrm{I}'_{9/2}\rangle = &-0.399|{}^4\mathrm{F}_{9/2}\rangle - 0.303|{}^2\mathrm{G}^1_{9/2}\rangle + 0.239|{}^2\mathrm{G}^2_{9/2}\rangle \\ &- 0.017|{}^4\mathrm{G}_{9/2}\rangle - 0.200|{}^2\mathrm{H}^1_{9/2}\rangle + 0.419|{}^2\mathrm{H}^2_{9/2}\rangle \\ &+ 0.690|{}^4\mathrm{I}_{9/2}\rangle\end{aligned}$$

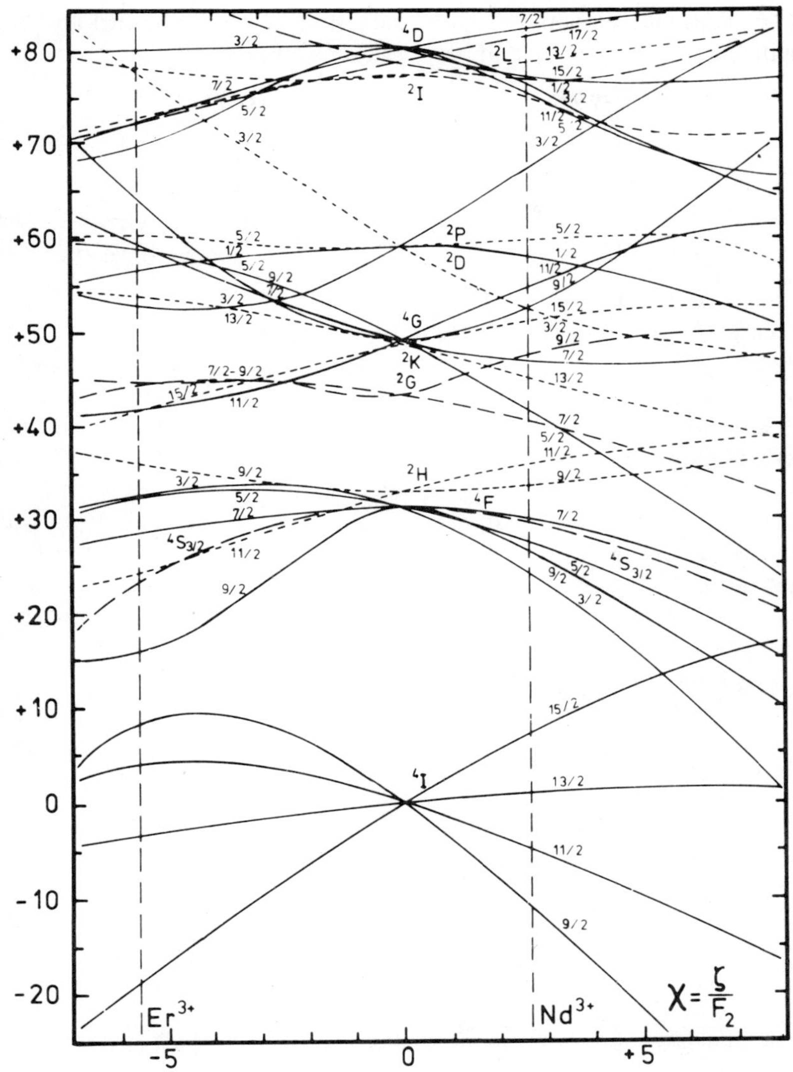

Fig. 4. The lowest terms of the $4f^3$ and $4f^{-3} = 4f^{11}$ configuration as a function of the normalized spin–orbit coupling constant $\chi = \zeta_{4f}/F_2$ [Di68].

The value of this constant where the best agreement with experimental data is achieved is also indicated for Er^{3+} and Nd^{3+}. The numbers on the various curves are the total angular momenta J of the terms. Energies at the ordinate are in units of F_2.

where the superscripts 1 and 2 distinguish the two ^2G and ^2H states for $4f^{-3}$. These wavefunctions can now be used to calculate various properties. For example the g_J-factor of a state is obtained by

$$g_J = \sum_i a_i^2 g_{Ji}(RS) \tag{2.25}$$

yielding for the $^4I'_{15/2}$ term in Er^{3+}

$$g_J(^4I'_{15/2}) = (0.984)^2 g_J(^4I_{15/2}) + (0.176)^2 g_J(^2K_{15/2}) + (0.019^2 g_J(^2L_{15/2}) = 1.196$$

instead of the pure Russell–Saunders value 1.20.

Table 3 compares a collection of g_J values obtained from crystal spectra [HeHü60] with those calculated in the Russell–Saunders and intermediate coupling approximations. For most terms agreement between the experimental g_J-factor and that calculated in the intermediate coupling approximation is quite good; thus the g_J-factor can be used to classify the states. The Russell–Saunders g_J-factors are in some instances markedly different. For example the $^4S'_{3/2}$ term in the Russell–Saunders approximation is $^4S_{3/2}$, a pure spin state, and therefore $g_J = 2$; by intermediate coupling, states with nonzero orbital angular momentum are mixed into it, resulting in $^4S'_{3/2}$, and since all the admixed terms have smaller g_J-factors, the g_J-factor of $^4S'_{3/2}$ is reduced to 1.68.

The interactions considered so far are the most important ones for describing RE free ion energy levels. In comparing experimental and theoretical results, however, discrepancies of the order of 100 cm^{-1}

TABLE 3

EXPERIMENTAL[a] AND THEORETICAL[b] g_J VALUES FOR Er^{3+}

Term	$g_{experimental}$	$g_{intermediate\ coupling}$	$g_{Russell–Saunders}$
$^4F_{9/2}$	1.12	1.117	1.33
$^4S_{3/2}$	1.65	1.68	2.00
$^2H_{11/2}$	1.13	1.133	1.09
$^4F_{7/2}$	1.23	1.212	1.20
$^4F_{5/2}$	1.05	1.053	1.02
$^4F_{3/2}$	0.70	0.77	0.40
$^2H_{9/2}$	1.00	1.076	0.91

[a] From [HeHü60].
[b] From [Ka61].

remain, and a number of workers have tried to reduce these discrepancies by taking into account further interactions. The most important one is the configuration interaction via the Coulomb interaction. It can be put into parameter form with three two-body integrals, usually denoted by α, β, γ [RaWy63], and six three-body integrals, usually denoted by T^2 to T^8 [Ju66a]. As far as the three-body interactions are concerned, the six terms represent only the dominant contributions, and higher order perturbation approximations would yield even more contributions. Then there are additional magnetic interactions, namely, the spin–spin interaction and the spin–other-orbit interaction, which are absorbed into three more parameters, M^0, M^2, and M^4. Finally because of the configuration interaction, the spin–orbit coupling can no longer be described by a single constant, and this effect can be put into three parameters, P^2, P^4, P^6

TABLE 4a

ELECTROSTATIC RADIAL INTEGRALS F_k AND SPIN–ORBIT COUPLING CONSTANTS OBTAINED FOR THE TRIVALENT LANTHANIDES IN VARIOUS MEDIA[a]

Matrix	F_2 (cm^{-1})	F_4 (cm^{-1})	F_6 (cm^{-1})	ζ (cm^{-1})
$Ce:LaCl_3$ [HeOr65]				625
Pr free ion	322	49.4	4.88	758
Pr aquo	305	46.2	4.43	740
Nd aquo	321	46.3	4.69	884
Pm aquo	338	49.8	4.62	1000
Sm aquo	364	56.7	5.42	1157
Eu aquo	369	56.2	5.64	1326
$GdCl_3 \cdot 6H_2O$	384	91.8	5.77	1450
Gd aquo	384	91.8	5.77	1450
Tb aquo	401	60.8	6.01	1709
Dy aquo	407	60.5	6.27	1932
Ho aquo	419	65.0	6.76	2141
Er free ion	441	68.3	7.49	2369
Er aquo	440	66.8	7.30	2380
Er $(C_2H_5SO_4)_3 \cdot 9H_2O$	462	70.7	7.80	2628
Tm aquo	461	70.7	7.80	2628
YbGaG [BuWi67]				2870

[a] All data except where explicitly noted are from [CaFi68a, CaFi68b].

TABLE 4b

ENERGY LEVEL PARAMETERS FOR $RE^{3+}:LaCl_3$ FROM FITS THAT INCLUDE HIGHER ORDER CORRECTIONS AND CRYSTAL FIELD INTERACTION[a,b]

	Pr	Nd	Pm	Sm	Eu	Gd	Tb	Dy	Ho	Er
F_2	304	319	337	347	375	379	405	412	424	436
F_4	45.9	47.9	49.9	52.2	55.4	55.5	59.1	60.3	61.7	64.0
F_6	4.45	4.82	5.27	5.45	5.65	6.09	5.83	6.19	6.35	6.67
α	22.9	22.1	21.0	21.6	16.8	(19)	17.5	17.2	17.2	15.9
β	−674	−650	−645	−724	(−640)	(−643)	(−630)	−622	−621	−632
γ	(1520)	1586	1425	(1700)	(1750)	1644	(1880)	1881	2092	(2017)
T^2	—	377	302	291	(370)	(315)	(340)	311	300	300
T^3	—	40	45	13	(40)	(44)	(40)	116	37	48
T^4	—	63	34	34	(40)	(40)	(40)	12	98	18
T^6	—	−292	−315	−193	(−330)	(−300)	(−330)	−474	−316	−342
T^7	—	358	554	288	(380)	(325)	(330)	413	440	214
T^8	—	354	(400)	330	(370)	(360)	(380)	315	372	449
ζ	744	880	1022	1168	1331	(1513)	1707	1920	2137	2370
M^{0c}	1.76	1.97	2.1	2.4	(2.38)	(2.82)	(3.00)	2.8	3.0	4.5
P^{2d}	275	255	319	341	245	495	590	591	523	667
$A_{20}\langle r^2\rangle$	54	82	72	93	95	108	143	97	108	108
$A_{40}\langle r^4\rangle$	−43	−42	−49	−34	−36	−34	−36	−41	−35	−34
$A_{60}\langle r^6\rangle$	−42	−45	−42	−39	−50	−43	−29	−29	−28	−26
$A_{66}\langle r^6\rangle$	444	438	426	446	499	450	286	272	279	268

[a] From [Cr77].
[b] Values are in wavenumbers. Values in parentheses were not freely varied.
[c] Only M^0 was varied; the ratios $M^2/M^0 = 0.56$, $M^4/M^0 = 0.38$ were maintained.
[d] Only P^2 was varied; the ratios $P^4/P^2 = 0.75$, $P^6/P^2 = 0.50$ were maintained.

TABLE 4c

ELECTROSTATICAL RADIAL INTEGRALS F_k AND SPIN–ORBIT COUPLING CONSTANT ζ CALCULATED IN THE RELATIVISTIC HARTREE–FOCK APPROXIMATION[a]

		F_2 (cm^{-1})	F_4 (cm^{-1})	F_6 (cm^{-1})	ζ (cm^{-1})
$4f^1$	Ce^{3+}	—	—	—	696.41
$4f^2$	Pr^{3+}	438	56.8	6.05	820.22
$4f^3$	Nd^{3+}	456	59.2	6.30	950.51
$4f^4$	Pm^{3+}	473	61.4	6.54	1091.46
$4f^5$	Sm^{3+}	489	63.5	6.76	1243.60
$4f^6$	Eu^{3+}	505	65.5	6.97	1407.71
$4f^7$	Gd^{3+}	520	67.5	7.18	1584.45
$4f^8$	Tb^{3+}	534	69.4	7.38	1774.46
$4f^9$	Dy^{3+}	549	71.2	7.58	1998.44
$4f^{10}$	Ho^{3+}	563	73.0	7.77	2197.06
$4f^{11}$	Er^{3+}	577	74.8	7.96	2431.00
$4f^{12}$	Tm^{3+}	590	76.5	8.15	2680.97
$4f^{13}$	Yb^{3+}	—	—	—	2947.69

[a] Calculation by Carnall *et al.* [CaCr77] in the approximation of [CoGr76].

[JuCr68, CrCr68]. Altogether, therefore we have 19 parameters with which to describe the free ion spectra, as compared to four in the naive approach outlined earlier. This naturally improves the quality of fit considerably and makes differences between theory and experiment practically disappear [CrCr76, CaCr76, CaCr77]. In terms of wavefunctions, however, this sophisticated approach produces only small changes as compared to those obtained from a basic intermediate coupling calculation, and therefore the intermediate coupling wavefunctions can generally be used with confidence and success.

Radial integrals F_k and the spin–orbit interaction constant ζ for all the REs are listed in Table 4a, from the work of Carnall *et al.* [CaFi68a, CaFi68b]. The additional parameters α, β, γ are incorporated into their fitting routine. However, they could only be determined with considerable uncertainty and are therefore not reproduced here. In these analyses, except for Ce^{3+} and Yb^{3+}, the crystal field interaction was neglected.

Carnall *et al.* [CaCr77] have analyzed the spectra of the Re^{3+}:$LaCl_3$ and of the odd electron RE^{3+}:LaF_3 systems by using all the mentioned parameters and by including also the crystal field interaction (see also

Chapter 3). This is of course the most complete and ultimately the only correct approach, and we list, therefore, the parameters obtained in this work in Table 4b. We notice, however, that the deviations of the F_k parameters and the spin–orbit coupling constants from those in Table 4a are not very large. Table 4c contains the F_k parameter and spin–orbit coupling constants calculated with a relativistic Hartree–Fock program [CaCr77]. A comparison with the experimental parameters (Tables 4a

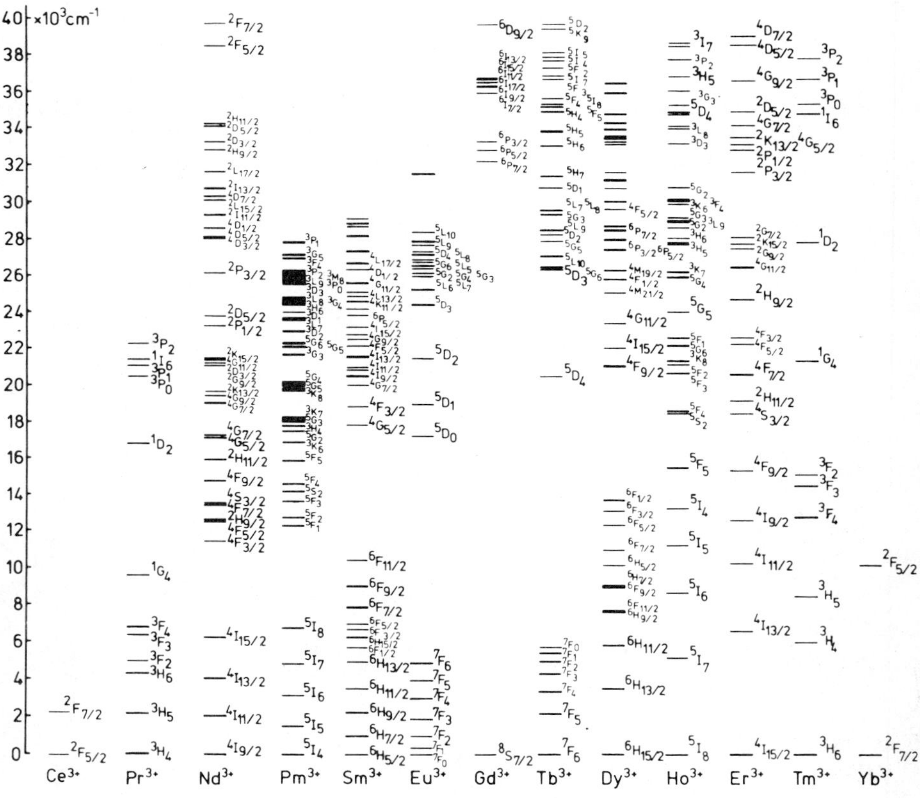

Fig. 5. Energy levels for RE^{3+}

The starting point for this diagram was the energy levels observed in $LaCl_3$ at Johns Hopkins University [Di68, DiCr63, CrMo67]. New and additional assignments have been added, but the diagram may not be complete. The classification is given in Russell–Saunders symbols, where that symbol has been chosen which applies for ζ (spin–orbit interaction) approaching zero. The original classification has been supplemented with that given by Carnall *et al.* [CaFi68b]. Data from other references have also been included: [CrCr76, CaCr76, HeHo61, CaCr77].

and 4b) shows that there are still substantial deviations between theory and experiment.

Finally, we state that for the trivalent REs in solids many of the energy levels below 30,000 cm^{-1} have already been analyzed, and for Ce^{3+}, Pr^{3+}, Nd^{3+}, Gd^{3+}, Er^{3+}, Tm^{3+}, and Yb^{3+} most of the observed levels up to 40,000 cm^{-1} have been classified. In Pm^{3+}, Sm^{3+}, Eu^{3+}, Tb^{3+}, Dy^{3+}, and Ho^{3+}, however, the free ion classification of some of the observed levels has not yet been established, even below 30,000 cm^{-1}. This does not mean a breakdown of the model employed so far. For these ions, however, there are very many free ion energy levels very close to each other, so that in many instances their crystal field levels overlap. Hence it is almost impossible to analyze the spectra in terms of discrete J states, which is why analysis of the ions mentioned has not progressed.

From the foregoing remarks it is evident that the energies that are the centers of gravity of the crystal field multiplets do not vary much for different hosts, meaning that their position is a "typical" host material is representative of that in all other crystals. A compilation of the energy levels found in the RE salts appears in Fig. 5. These data were collected primarily at the Johns Hopkins University [Di68], but many of the classifications had also been obtained at other laboratories in other hosts. This figure is a very useful guide for the analysis of an unknown spectrum;

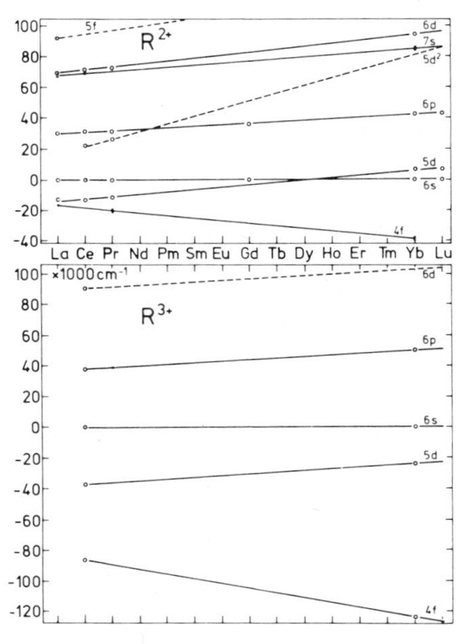

Fig. 6. Approximate relative position of the $4f^N$ configurations and the next excited configurations for the trivalent and divalent RE ions [Di68, DiCr63]. The solid and broken lines were drawn to connect the data points.

if one is interested in a probe with special properties it may also be of some help. In constructing Fig. 5 from the Dieke's original version [Di68], we added a few new classifications, provided by Carnall *et al.* [CaFi68b]. The work of Crosswhite *et al.* [CrCr76, CaCr76, CaCr77] was also used and the $^6F_{1/2}$ term for Dy^{3+} is taken from the work of Hellwege [HeHo61].

The single most intensively studied system for classifying RE energy levels is that of Carnall *et al.* [CaFi68b] on REs in aqueous solution. Since the environments of RE ions in solutions do not differ greatly from those in solids, this work may also be useful for analyzing crystal spectra.

Figure 6 is a result of the analysis of the free ion spectra of some trivalent and divalent ions; it shows the approximate positions of the next excited configurations with respect to the $4f^N$ configuration [Di68]. For the trivalent ions we see that the $4f^N$ configuration is relatively isolated and that the next excited configuration is $4f^{N-1}5d$, which starts about 50,000 cm^{-1}

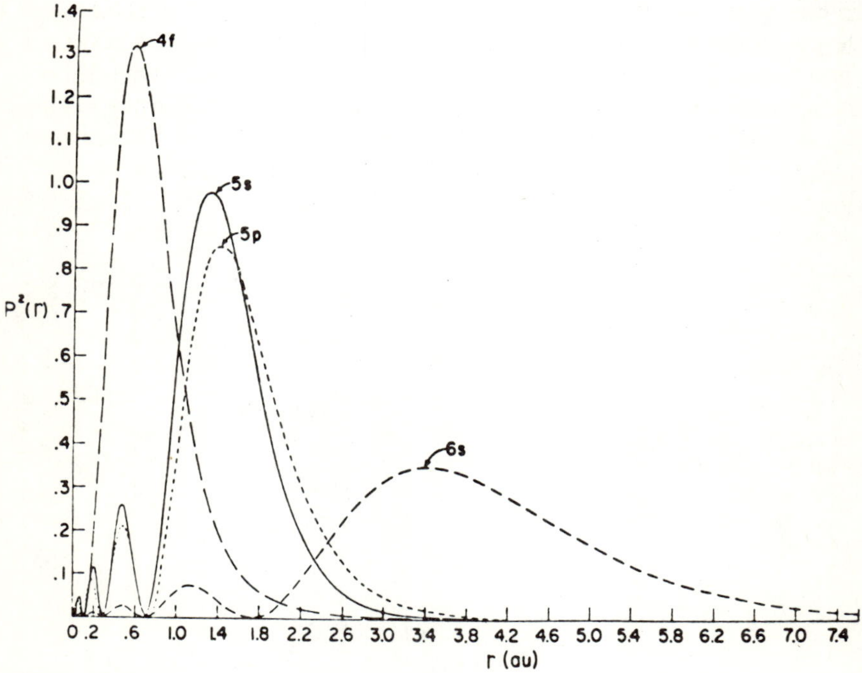

Fig. 7. Square of the radial wavefunction times the squared radius for the 4f, 5s, 5p, and 6s electrons in Gd^+ [FrWa62]. For RE^{3+} ions this picture will be contracted somewhat because of the larger effective nuclear charge.

above the ground term of the $4f^N$ configuration for Ce^{3+} and about 100,000 cm^{-1} above it for Yb^{3+}; these data are for the free ions; the respective figures for the solid seem somewhat (about 30%) lower, but the general trend is preserved.

The $4f^N \to 4f^{N-1}5d$ transitions are very intense because they are no longer forbidden by the parity selection rule for electric dipole radiation. Where they occur, they may therefore obscure $4f^N \to 4f^{N'}$ transitions. This does not happen for RE^{3+} ions in the visible spectral region. For RE^{2+} ions, however, the $4f^N \to 4f^{N-1}5d$ energy separation is smaller and for some ions in the visible region leads to very intense $4f^N \to 4f^{N-1}5d$ transitions, which mask the $4f^N \to 4f^{N'}$ transitions.

Figure 7 shows the radial distribution functions of the 4f, 5s, 5p, and 6s electrons for Gd^+ as obtained from Hartree–Fock calculation by Freeman and Watson [FrWa62]. This graph shows that the 4f electrons are indeed inner electrons with relatively small $\langle r^n \rangle$ values. As we shall see, the crystal field interaction is proportional to these $\langle r^n \rangle$ values and therefore this interaction will be small in REs. It is also evident that the 4f wavefunctions do not extend very far beyond the 5s and 5p shells. Therefore, all quantities of a solid that depend on an overlap of the 4f wavefunctions with those of a neighboring ion (e.g., direct exchange) have to be small in RE compounds.

We conclude this section with a word on notation: So far we have distinguished the pure Russell–Saunders term (e.g., ${}^4I_{15/2}$) from the intermediate coupling term by using the primed symbol ${}^4I'_{15/2}$. There are unfortunately two ways to ascribe a Russell–Saunders symbol to a specific term in the intermediate coupling case. We can use the Russell–Saunders symbol for the wavefunction that gives the largest contribution to the total wavefunction of the term (in Russell–Saunders eigenfunctions, of course) or we can use the symbol for the Russell–Saunders term into which the intermediate coupling term collapses for vanishing spin–orbit coupling. In most cases the two methods lead to the same labeling. In the few cases where they do not, we shall follow the second convention [Ka61] and again use unprimed symbols.

CHAPTER 3

Trivalent Ions in the Static Crystal Field

3.1 Introduction

The 4f shell of RE ions is a nonclosed shell and therefore has an aspherical charge distribution. If the ion is introduced into a crystal, the ion experiences an inhomogeneous electrostatic field, the so-called crystal field, which is produced by the charge distribution in the crystal. This crystal field distorts the closed shells of the RE ion. The effect on the nonclosed 4f shell is yet more important: It removes to a certain degree (which depends on the crystal symmetry), the M_J degeneracy of the free ion 4f levels discussed in Chapter 2, thus producing a major modification of the energy level scheme. Experiments have shown that for trivalent RE ions the crystal field interaction (i.e., the splitting of the free ion terms produced by placing the ion in a crystal) is smaller than the energy separation of the various free ion terms. This means that an understanding of this crystal field splitting can be obtained by applying the potential produced by the crystal on the free ion 4f wavefunctions.

The principal features of the phenomenon can now be illustrated by a simple example. Let us assume a free ion energy level with total angular momentum $J = \frac{3}{2}$, such as the $^4S_{3/2}$ term of Er^{3+} at $\sim 18{,}000\ cm^{-1}$ (see Fig. 8). Note that in the intermediate coupling approximation this term has the wavefunction

$$|^4S'_{3/2}\rangle = 0.8|^4S_{3/2}\rangle - 0.4|^2P_{3/2}\rangle - 0.3|^2D^1_{3/2}\rangle + 0.2|^4F_{3/2}\rangle + \cdots$$

where the dots indicate smaller contributions [Ka61]. This shows that its total angular momentum is not a pure spin momentum but also contains sizeable orbital contributions. Since the crystal field only acts on the orbital part of the 4f wavefunctions, the mixed-in orbital contributions are responsible for a nonzero crystal field splitting of the $^4S_{3/2}$ term.

The spectrum in Fig. 8 also serves as an example of how a high resolution absorption spectrum of an RE ion looks [HeHü60] and of the kind of information that can be obtained from its analysis. The figure shows the $^4I_{15/2} \rightarrow {}^4S_{3/2}$ transition of Er^{3+} in erbium ethylsulfate ($Er(C_2H_5SO_4)_3{\cdot}9H_2O$), a hexagonal crystal whose crystal field potential contains a contribution that is able to split a $J = \frac{3}{2}$ free ion term. The energy separation between the various crystal field levels into which a

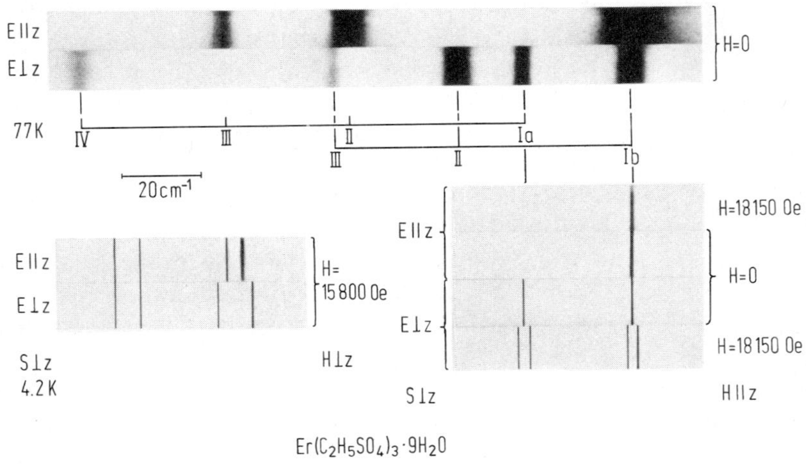

Fig. 8. Absorption spectrum of the $^4I_{15/2} \rightarrow {}^4S_{3/2}$ transition in $Er(C_2H_5SO_4)_3{\cdot}9H_2O$. **E** is the electric vector of the radiation, **S** the point vector (propagation direction of the light), H the external applied magnetic field, and z the hexagonal crystal axis. Lower right: $Er^{3+}_{0.1}: Y(C_2H_5SO_4)_3 \cdot 9H_2O$, $T = 4.2$ K. The two middle spectra are for $H = 0$ and different polarizations; the two outer spectra show the effect of an applied field parallel to the hexagonal crystal axis. Lower left: same as outer two spectra at the lower right, but magnetic field applied perpendicular to the crystal axis. Upper: $Er(C_2H_5SO_4)_3{\cdot}9H_2O$, $T = 85$ K. The elevated temperature populated excited crystal field levels of the ground term, which give rise to additional lines; a and b label the crystal field levels of the $^4S_{3/2}$ excited term, whereas those of the ground term $^4I_{15/2}$ are labeled with Roman numbers [HeHü60].

free ion term is split after the RE ion has been incorporated into the crystal is of the order of 50 cm^{-1} ($\approx$ 75 kT). At 4.2 K, therefore, only the lowest crystal field level of the ground term is populated (because the separation between the first and second crystal field levels of the ground term is large compared to the thermal energy corresponding to 4.2 K). Then the absorption spectrum $^4I_{15/2} \rightarrow {}^4S_{3/2}$ gives, via the experimental splitting patterns, the crystal field splitting of the excited $^4S_{3/2}$ term. This spectrum is shown at the lower right in Fig. 8 ($H = 0$, H being the external magnetic field). The figure also shows that these spectra are polarized; the reason for this will be outlined later. Finally, we see that the absorption lines can easily be split by a magnetic field of modest size. This is a proof that the crystal field levels are doublets. Analysis of this splitting yields the magnetic splitting factors of the ground and excited crystal field levels. The two spectra to the left of those just discussed show the effect of a magnetic field perpendicular to the crystal axis (at the right the magnetic field is parallel to the axis). This splitting pattern, though different, can also be analyzed. Finally, at the top of Fig. 8 we see the effect on the spectra of elevated temperatures (e.g., 77 K, which is comparable to the crystal field energy). Additional transitions, which originate from the excited crystal field levels of the ground term are evident, and the spectra look more complicated. On the other hand, these spectra can be used to determine the crystal field energies of the ground term.

We shall now try to understand the observed behavior of the $^4S_{3/2}$ term (namely, the 4.2 K absorption spectra) in a qualitative way. To do so we draw on our treatment of the Coulomb interaction of the 4f electrons (eqs. (2.3)–(2.5), Section 2.1). The crystal field interaction comes about by the interaction of the 4f electrons with all the charges of the crystal; except for trivial factors, it is therefore given by one over the radius vector between the 4f electrons and the crystal charges. This function is given, apart from radial factors, by the product of two spherical harmonics Y_{kq}, one containing the coordinates of the 4f electrons, the other containing coordinates of a crystal charge. The latter has to be integrated over the whole crystal, whereupon it gives the strength of the crystal potential at the site of the RE ion. We also note that only even k occur because we are dealing with an energy.

In the case of $Er(C_2H_5SO_4)_3{\cdot}9H_2O$, the crystallographic data tell us that the Er ion is located at a site of hexagonal symmetry [Ke37, FiRu59]. Then the development of $1/\mathbf{r}$ into spherical harmonics yields, for the crystal part of the product that gives the interaction energy, contributions Y_{20}, Y_{40}, ... We are interested in the interaction of a term having $J = \frac{3}{2}$ ($^4S_{3/2}$) with this potential. The charge distribution of a $J = \frac{3}{2}$ term contains only spherical harmonics Y_{20} and $Y_{2,\pm 2}$, where only the first is of interest because due to the crystal symmetry only it sees a field from the crystal. Therefore we have, for the interaction energy of the $J = \frac{3}{2}$ term with the crystal field (if we absorb all constants into a prefactor K_{20}),

$$W(J = \tfrac{3}{2}) = K_{20} \cdot Y_{20}(\text{crystal}) \cdot Y_{20}\,(4f) \qquad (3.1a)$$

We notice that $Y_{20}\,(4f) \sim (3M_J^2 - J(J+1))$, where M_J is the directional quantum number of J and therefore

$$W(J = \tfrac{3}{2}) = \bar{K}_{20}\,Y_{20}(\text{crystal}) \cdot (3M_J^2 - J(J+1)) \qquad (3.1b)$$

where $\bar{K}_{20}$ now also contains the radial integral over the 4f wavefunction. This formula shows that if $Y_{20}(\text{crystal}) \neq 0$, the $^4S_{3/2}$ free ion term is split into two levels, characterized by $M_J = \pm\frac{1}{2}$ and $\pm\frac{3}{2}$, respectively: The M_J degeneracy of the free ion term is removed partly by the crystal field, which means that the $^4S_{3/2}$ free ion term is split into two crystal field levels, as shown in Fig. 8. We see that in systems with an odd number of 4f electrons the degeneracy is not removed completely but remains with respect to $\pm M_J$; this is called the Kramers degeneracy.

Formula (3.1) indicates that for ions with an even number of 4f electrons, which produce states with integral numbers of J, not all states must be doubly degenerate because at least the state with $M_J = 0$ can be a singlet.

This simple example shows the kind of information that can be obtained from analyzing RE crystal spectra. The existence of the splitting tells us that the development of the crystal potential into spherical harmonics contains a second order term (which means that the crystal symmetry must be lower than cubic), and from the size of the splitting the size of the crystal field at the site of the 4f electrons can be deduced.

3.2 Parametrization of the Crystal Field Splitting

Crystal field splitting has two aspects: one concerns the symmetry of the problem, namely, the number of levels into which an ion's free ion J terms are split in a crystal field of a given symmetry; the second concerns the actual size of the crystal field splitting. Whereas the first topic is now completely understood, the second is being actively researched, and the various mechanisms that determine the magnitude of the crystal field splitting are by no means completely understood (see, e.g., [Ne71]).

Therefore, a parametrization scheme has been developed that makes it possible to absorb a measured crystal field spectrum into a few parameters, whose meaning we will now discuss.

In the most naive picture, a crystal can be visualized as built up of point charges at the crystallographic positions of the ions. The potential at the site of an RE ion can then be calculated and used to determine the crystal field splitting of the RE free ion terms. This approach, called the point charge model, does give the qualitative features of the problem, but actual calculations show that the magnitude of the crystal field splitting is underestimated in this scheme by up to an order of magnitude. At best, then, the point charge model can be used for illustrative purposes but not for a quantitative description of the crystal field interaction. A realistic description of this interaction must take into account that the crystal is built up of the spatially extended charge clouds of the individual ions. These charge clouds can penetrate each other and thereby interact. The model we will use to describe the crystal field interaction contains the following elements.

(1) The ions are considered static in the crystal (lattice vibrations and their effect on the energy levels are neglected).

(2) The 4f electrons of one RE ion are regarded as representative of those of all the RE ions in the crystal and the interaction of 4f electrons of adjacent RE ions is neglected.

(3) The crystal consists of extended charge distributions. This produces an overlap of the charge distributions of the neighboring ions and the 4f electrons. In addition, a charge transfer between the 4f electrons and the electrons of the ligands can take place. Both effects contribute to the crystal field interaction.

(4) The 4f electrons of one ion are considered to be independent of each other; that is correlation effects play no significant role.

The first statement is trivial as long as we are not considering problems that explicitly depend on the interaction of the phonons with the RE ions, such as linewidth and temperature-dependent line position. But some crystals (e.g., the RE vanadates, arsenates, and phosphates) undergo crystallographic changes as a function of temperature, changes that are driven by the phonons (Jahn–Teller effect). These effects, which we neglect for the moment, are treated in Chapter 8.

The second item means that interactions between the 4f electrons of different ions are negligible. This seems a reasonable approximation in view of the electronic structure of the RE ions and is equivalent to saying that the RE energy levels have no dispersion.† This is true only for insulators (but not for semiconductors and metals).

The third element stems from the realization that even for RE compounds the point charge model is a very poor approximation. For an actual calculation a molecular orbital type of procedure is necessary.

The last point is necessary in order to make the problem mathematically tractable. Also, there is no evidence so far that this assumption introduces gross errors.

With the foregoing assumptions, we calculate the crystal potential $\phi(r_i, \varphi_i, \vartheta_i)$ at the site of the 4f electrons and the potential energy of the 4f electrons in this potential. If the crystal has charge density $\rho(\mathbf{R})$ and the 4f electrons have radius r_i, we obtain (using eqs. (2.3a), (2.3b), (2.4), (2.5))

$$V = -\sum_i \int \frac{e_i \rho(\mathbf{R})}{|\mathbf{R} - \mathbf{r}_i|} d\tau = -\sum_{k,i} e_i \int \rho(\mathbf{R}) P_k(\cos(\mathbf{R}, \mathbf{r}_i)) \frac{r_<^k}{r_>^{k+1}} d\tau$$

where $r_<$ and $r_>$ are, respectively, the smaller and larger value of $\mathbf{R}$ and $\mathbf{r}_i$.

$$\begin{aligned} V &= -\sum_{k,q,i} e_i \int \rho(\mathbf{R}) \frac{4\pi}{2k+1} (-1)^q Y_{kq}(\vartheta_i, \varphi_i) Y_{k-q}(\vartheta, \varphi) \frac{r_<^k}{r_>^{k+1}} d\tau \\ &= -\sum_{k,q,i} e_i \int (-1)^q \rho(\mathbf{R}) C_{kq}(\vartheta_i, \varphi_i) C_{k-q}(\vartheta, \varphi) \frac{r_<^k}{r_>^{k+1}} d\tau \\ &\equiv \sum_{k,q,i} B_{kq} C_{kq}(\vartheta_i, \varphi_i) \end{aligned} \tag{3.2}$$

where the crystal field parameter B_{kq} is defined as

$$B_{kq} = -e \int (-1)^q \rho(\mathbf{R}) C_{k-q}(\vartheta, \varphi) \frac{r_<^k}{r_>^{k+1}} d\tau$$

(The notation is that of Wybourne [Wy65].)

To calculate the matrix elements of V we apply the rules outlined in Section 2.2 for calculating the matrix elements of tensor operators. All the radial integrals that are hard to obtain exactly are absorbed (together with the effect of the crystal potential) into the crystal field parameters B_{kq}.

In early treatments of the crystal field interaction it was assumed that in RE compounds the point charge contribution would be the dominant part of the crystal field interaction. Then in eq. (3.2) the integral over the lattice can be replaced by a sum over all lattice points and $r_<^k$ can be replaced by r_i^k. The latter replacement can be performed as long as the

† See footnote on p. 3 and [CoMe75].

charge distribution of the crystal does not enter that of the 4f electrons (as long as $r_i < R$), which implies that the potential acting on the 4f electrons obeys the Laplace equation ($\Delta\phi(r_i, \varphi_i, \vartheta_i) = 0$) at the position of the 4f electrons. This has led to slightly different notation, which we mention, because it is used so frequently. Since these symbols originate from a Hamiltonian that was derived from a point charge model, the radial $\langle r^k \rangle$ integrals for the 4f electrons show up explicitly. Abragam and Bleaney [AbBl70] use the Hamiltonian

$$V = \sum_{k,q,i} \bar{B}_{kq} r_i^k Y_{kq}(\vartheta_i, \varphi_i) \tag{3.3}$$

which yields crystal field parameters $\bar{B}_{kq}\langle r^k \rangle$ with

$$\bar{B}_{kq} = -e \int \frac{\rho(\mathbf{R})}{R^{k+1}} (-1)^q C_{k-q}(\vartheta, \varphi)\, d\tau \tag{3.4}$$

Then Stevens [Ste52] in his original work on crystal field parametrization used the potential energy form

$$V = \sum_{k,q,i} A_{kq} P_{kq}(x_i, y_i, z_i), \qquad q \geq 0 \tag{3.5}$$

where the P_{kq} are polynomials given in Table 5. Here the calculation of the matrix elements again results in a factor $\langle r^k \rangle$ and crystal field parameters $A_{kq}\langle r^k \rangle$ are obtained. These $A_{kq}\langle r^k \rangle$ parameters are the ones most often used in the literature and we shall use them, too. Tables 6a and 6b relate

TABLE 5

POLYNOMIALS $P_{kq}(x, y, z)$ OF EVEN DEGREE THAT ARE USED IN EQ. (3.5)[a]

$P_{20} = 3z^2 - r^2$

$P_{22} = x^2 - y^2$

$P_{40} = 35z^4 - 30r^2z^2 + 3r^4$

$P_{42} = (7z^2 - r^2)(x^2 - y^2)$

$P_{43} = xz(x^2 - 3y^2)$

$P_{44} = x^4 - 6x^2y^2 + y^4$

$P_{60} = 231z^6 - 315r^2z^4 + 105r^4z^2 - 5r^6$

$P_{62} = (16z^4 - 16(x^2 + y^2)z^2 + (x^2 + y^2)^2)(x^2 - y^2)$

$P_{63} = (11z^2 - 3r^2)(x^2 - 3y^2)xz$

$P_{64} = (11z^2 - r^2)(x^4 - 6x^2y^2 + y^4)$

$P_{66} = x^6 - 15x^4y^2 + 15x^2y^4 - y^6$

[a] From [AbBl70].

TABLE 6a

RELATIONS BETWEEN THE $A_{kq}\langle r^k\rangle$ OF EQ. (3.5) AND THE $\bar{B}_{kq}\langle r^k\rangle$ OF EQ. (3.3)[a]

$A_{20} = \frac{1}{(2\pi)^{1/2}}(5/8)^{1/2}\bar{B}_{20}$	$A_{60} = \frac{1}{(2\pi)^{1/2}}\frac{(26)^{1/2}}{32}\bar{B}_{60}$
$A_{22} = \frac{1}{(2\pi)^{1/2}}(15/4)^{1/2}\bar{B}_{22}$	$A_{62} = \frac{1}{(2\pi)^{1/2}}\frac{(105)^{1/2}(26)^{1/2}}{32}\bar{B}_{62}$
$A_{40} = \frac{1}{(2\pi)^{1/2}}\frac{3(2)^{1/2}}{16}\bar{B}_{40}$	$A_{63} = \frac{1}{(2\pi)^{1/2}}\frac{(26 \times 105)^{1/2}}{16}\bar{B}_{63}$
$A_{42} = \frac{1}{(2\pi)^{1/2}}\frac{3(5)^{1/2}}{4}\bar{B}_{42}$	$A_{64} = \frac{1}{(2\pi)^{1/2}}\frac{3(13 \times 28)^{1/2}}{32}\bar{B}_{64}$
$A_{43} = \frac{1}{(2\pi)^{1/2}}\frac{3(70)^{1/2}}{4}\bar{B}_{43}$	$A_{66} = \frac{1}{(2\pi)^{1/2}}\frac{(13 \times 21 \times 22)^{1/2}}{32}\bar{B}_{66}$
$A_{44} = \frac{1}{(2\pi)^{1/2}}\frac{3(35)^{1/2}}{8}\bar{B}_{44}$	

[a] From [AbBl70].

TABLE 6b

RELATIONS BETWEEN THE $A_{kq}\langle r^k\rangle$ OF EQ. (3.5) AND THE B_{kq} OF EQ. (3.2)[a]

$B_{20} = 2A_{20}\langle r^2\rangle$	$B_{60} = 16A_{60}\langle r^6\rangle$
$B_{22} = \frac{(6)^{1/2}}{3}A_{22}\langle r^2\rangle$	$B_{62} = \frac{16(105)^{1/2}}{105}A_{62}\langle r^6\rangle$
$B_{40} = 8A_{40}\langle r^4\rangle$	$B_{63} = \frac{-8(105)^{1/2}}{105}A_{63}\langle r^6\rangle$
$B_{42} = \frac{2(10)^{1/2}}{5}A_{42}\langle r^4\rangle$	$B_{64} = \frac{8(14)^{1/2}}{21}A_{64}\langle r^4\rangle$
$B_{43} = \frac{-2(35)^{1/2}}{35}A_{43}\langle r^4\rangle$	$B_{66} = \frac{16(231)^{1/2}}{231}A_{66}\langle r^6\rangle$
$B_{44} = \frac{4(70)^{1/2}}{35}A_{44}\langle r^4\rangle$	

[a] From [Ka70].

the various definitions of the parameters. The notation $\bar{B}_{kq}\langle r^k\rangle$ and $A_{kq}\langle r^k\rangle$ unfortunately gives the impression that the radial integrals can be extracted from these parameters. Since there is charge exchange involving the 4f electrons, this is not the case; therefore, we emphasize that experiments always give the products $\bar{B}_{kq}\langle r^k\rangle$ and $A_{kq}\langle r^k\rangle$, and there is no way to separate them.†

We shall now deal with the calculation of the angular part of the crystal field matrix elements. If eq. (3.2) is used as a starting point, evaluating them is easy because again the tensor operator technique can be used to obtain

$$
\begin{aligned}
&(\mathrm{f}^N SLJM_J|V|\mathrm{f}^N S'L'J'M'_J)\\
&\quad = \sum_{kq} B_{kq}(\mathrm{f}^N SLJM_J|U_{kq}|\mathrm{f}^N SL'J'M'_J)(\mathrm{f}\|C_k\|\mathrm{f})
\end{aligned}
\tag{3.6}
$$

(Since V is an electrical interaction, it cannot act on the spin and therefore $S = S'$.) Here we have used the normalized U_{kq} in order to adhere to the usual nomenclature (S and L couple to J; V acts only on part 2, L). To get the proper normalization $((\mathrm{f}\|U_k\|\mathrm{f}) = 1)$ we have to multiply by $((\mathrm{f}\|C_k\|\mathrm{f}))$; with eq. (2.9) and (2.17b) this yields

$$
\begin{aligned}
&(\mathrm{f}^N SLJM_J|U_{kq}|\mathrm{f}^N SL'J'M'_J)\\
&\quad = (-1)^{J-M_J}\begin{pmatrix} J & k & J' \\ -M_J & q & M'_J \end{pmatrix}(\mathrm{f}^N SLJ\|U_k\|\mathrm{f}^N SL'J')
\end{aligned}
\tag{3.7a}
$$

and

$$
\begin{aligned}
(\mathrm{f}^N SLJ\|U_k\|\mathrm{f}^N SL'J') &= (-1)^{S+L'+J+k}\{(2J+1)(2J'+1)\}^{1/2}\\
&\quad \times \begin{Bmatrix} J & J' & k \\ L' & L & S \end{Bmatrix}(\mathrm{f}^N SL\|U_k\|\mathrm{f}^N SL')
\end{aligned}
\tag{3.7b}
$$

where the matrix elements in (3.7b) can be found in the tables of Nielson and Koster [NiKo64] and the $3j$ and $6j$ symbols may be obtained from the tables of Rotenberg *et al.* [RoBi59]. The reduced matrix elements in (3.7a) have been given for many RE ion terms by Dieke [Di68] in the intermediate coupling approximation.

† The situation with respect to the crystal field parameters is further complicated because the Johns Hopkins Group has frequently used the definition $B_{kq} = A_{kq}\langle r^k\rangle$ [where the $A_{kq}\langle r^k\rangle$ are those of eq. (3.5)] [Di68], and recently Crosswhite [CrCr76] has used the notation $C_{kq} = B_{kq}$ [where the B_{kq} are those of eq. (3.2)].

These formulas are the most general ones. The matrix elements are diagonal only in S, and the formulas make it possible to evaluate matrix elements of the crystal field Hamiltonian between states of different J as well, which can be important for close-lying free ion terms.

We now discuss the actual handling of the problem in more detail. First, we determine which terms (k, q) of the series in eq. (3.2) are nonzero. The triangular condition for the evaluation of matrix elements (which states that if in a matrix element an operator of rank k connects angular momenta l_1 and l_2 so that the matrix element is nonzero, the condition $l_1 + l_2 \leq k$ must hold) shows that for f electrons $k \leq 6$ always. In addition, since the crystal field splitting is an energy, matrix elements have to be real and hence the condition that k is even holds. The number of nonzero terms is then further restricted by the point symmetry at the site of the RE ion of the crystal. If we take a common example, the D_{3h} symmetry, which means there is a threefold axis, a twofold axis perpendicular to it, and a reflection plane perpendicular to the threefold axis which makes B_{66} real (a crystal symmetry that is found approximately in the RE ethylsulfates), the development can be written

$$V = B_{20}C_{20} + B_{40}C_{40} + B_{60}C_{60} + B_{66}(C_{66} + C_{6\,-6}) \tag{3.8}$$

where for convenience the summation over i, the number of 4f electrons, has been dropped.

Since many of the examples in this book deal with the ethylsulfates, the most important part of their crystal structure is shown in Fig. 9 [Ke37, FiRu59]. The immediate environment of an RE ion, which we assume to be responsible for the crystal field the ion experiences, consists of nine water molecules. Six are deployed at the corners of a triangular prism, at the center of which is the RE ion, with three of the water molecules above and three below it. Three more water molecules are positioned in the plane of the RE ion, turned about 60° with respect to the other six water molecules. If the center of a polar coordinate system is superimposed on the center of the RE ion, the oxygen ions of the water molecules have the positions (the values are for Er):

(a) three oxygens at (R_1, θ_1, ϕ_1), where $R_1 = 2.37$ Å, $\theta_1 = 45°$, $\phi_1 = \frac{1}{3}\pi, \pi, \frac{5}{3}\pi$

(b) three oxygens at $(R_1, \pi - \theta_1, \phi_1)$;

(c) three oxygens at (R_2, θ_2, ϕ_2), where $R_2 = 2.52$ Å, $\theta_2 = \pi/2$, $\phi_2 = -\alpha, \frac{2}{3}\pi - \alpha, \frac{4}{3}\pi - \alpha$;

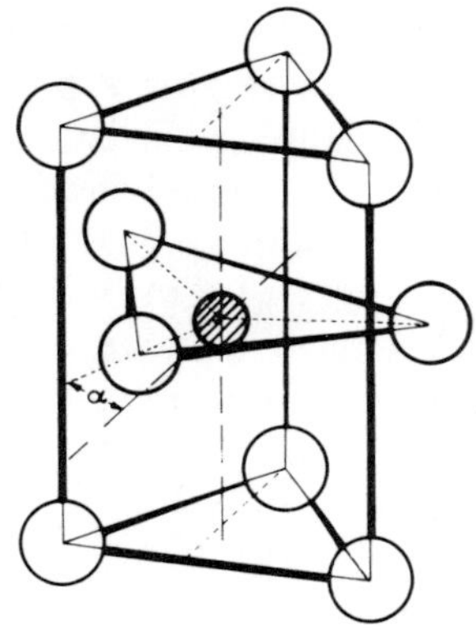

Fig. 9. Approximate position of the nine nearest neighbor water molecules surrounding an RE^{3+} ion in the ethylsulfate lattice ($Er(C_2H_5SO_4)_3 \cdot 9H_2O$).
▨ is the rare earth ion; ○ is a water molecule. The RE^{3+} ion is taken as the center of a polar coordinate system. The middle three water molecules and the RE^{3+} ion lie in a plane and α is the angle by which the middle triangle of water molecules is turned in the azimuthal plane with respect to the upper and lower triangle [Or61].

and $\alpha = 4°23'$, $5°5'$, and $6°25'$ for Er, Y, and Pr ethylsulfate, respectively. The exact point symmetry at the site of the RE ion is C_{3h}. For $\alpha = 0$, the point symmetry is higher, namely D_{3h}. Spectroscopic measurements (see, e.g., [GrHe61]) show that the point symmetry can indeed be treated as D_{3h}, so we shall do so here. For C_{3h} symmetry the Hamiltonian in eq. (3.8) has to be augmented by an imaginary term $iB_{66}(C_{66} + C_{6-6})$. Since for the ethylsulfates (and also for the tribromides and trichlorides) the deviation from D_{3h} symmetry is small, this term is usually neglected.

At present it is almost impossible to determine the B_{kq} in eq. (3.2) from actual crystallographic data. The Hamiltonian (3.2) is therefore used to parametrize the observed crystal field splitting in terms of the parameters B_{kq}. If a great number of crystal field energy levels has been determined experimentally, these energy levels are analyzed by evaluating the matrix elements of C_{kq} and treating the B_{kq} as adjustable parameters. The crystal field splittings of many different RE compounds have been analyzed in this way. The accuracy with which a measured spectrum can be described in terms of only a few parameters is generally quite good (of the order of $10\ cm^{-1}$ or less).

Using eq. (3.6) is very tedious for order of magnitude calculations. Also, in cases where one free ion term is well separated from the next (which is the case in most instances for the ground terms), the matrix elements between different J states need not to be taken into account. Then only the diagonal terms (in J) have to be calculated. We first introduce the nomenclature for historical reasons, which is slightly different. The matrix elements of the crystal field parameters within one J term are obtained by using the so-called Stevens operator equivalent factors [Ste52, ElSt52, ElSt53a, ElSt53b]. The crystal field Hamiltonian is written in the form of eq. (3.5), the first term of which is

$$V_{20} = \sum_i A_{20}(3z_i^2 - r_i^2) \tag{3.9}$$

and in order to obtain the crystal field energy we have to calculate the matrix element

$$\left(\mathrm{f}^N SLJM_J \middle| \sum_i A_{20}(3z_i^2 - r_i^2) \middle| \mathrm{f}^N SLJM'_J \right) \tag{3.10}$$

This matrix element can be related to the one of the operator $(3J_z^2 - J(J+1))$ (because both $(3z^2 - r^2)$ and $(3J_z^2 - J(J+1))$ have

TABLE 7a

EQUIVALENT OPERATORS FOR THE POLYNOMIALS IN TABLE 5[a, b]

$k = 2$	$O_{20} = 3J_z^2 - J(J+1)$
	$O_{22} = \frac{1}{2}(J_+^2 + J_-^2)$
$k = 4$	$O_{40} = 35J_z^4 - 30J(J+1)J_z^2 + 25J_z^2 - 6J(J+1) + 3J^2(J+1)^2$
	$O_{42} = \frac{1}{2}\{(7J_z^2 - J(J+1) - 5)(J_+^2 + J_-^2)\}_s$
	$O_{43} = \frac{1}{2}\{(J_z(J_+^3 + J_-^3)\}_s$
	$O_{44} = \frac{1}{2}(J_+^4 + J_-^4)$
$k = 6$	$O_{60} = 231J_z^6 - 315J(J+1)J_z^4 + 735J_z^4 + 105J^2(J+1)^2J_z^2 - 525J(J+1)J_z^2 + 294J_z^2 - 5J^3(J+1)^3 + 40J^2(J+1)^2 - 60J(J+1)$
	$O_{62} = \frac{1}{2}\{[33J_z^4(18J(J+1) + 123)J_z^2 + J^2(J+1)^2 + 10J(J+1) - 102](J_+^2 + J_-^2)\}_s$
	$O_{63} = \frac{1}{2}\{(11J_z^3 - 3J(J+1)J_z - 59J_z)(J_+^3 + J_-^3)\}_s$
	$O_{64} = \frac{1}{2}\{(11J_z^2 - J(J+1) - 38)(J_+^4 + J_-^4)\}_s$
	$O_{66} = \frac{1}{2}(J_+^6 + J_-^6)$

[a] From [AbBl70].
[b] The notation $\{A, B\}_s$ is for $\frac{1}{2}(AB + BA)$.

TABLE 7b

MATRIX ELEMENTS OF THE OPERATORS O_{kp} IN TABLE 7a[a,b]

J	F	M_J: $\pm 1/2$	$\pm 3/2$	$\pm 5/2$	$\pm 7/2$	$\pm 9/2$	$\pm 11/2$	$\pm 13/2$	$\pm 15/2$
Matrix elements of $O_{20} = 3J_z^2 - J(J+1)$									
3/2	3	−1	1						
5/2	2	−4	−1	5					
7/2	3	−5	−3	1	7				
9/2	6	−4	−3	−1	2	6			
11/2	1	−35	−29	−17	1	25	55		
13/2	6	−8	−7	−5	−2	2	7	13	
15/2	3	−21	−19	−15	−9	−1	9	21	35
Matrix elements of $O_{40} = 35J_z^4 - 30J(J+1)J_z^2 + 25J_z^2 - 6J(J+1) + 3J^2(J+1)^2$									
5/2	60	2	−3	1					
7/2	60	9	−3	−13	7				
9/2	84	18	3	−17	−22	18			
11/2	120	28	12	−13	−33	−27	33		
13/2	60	108	63	−13	−92	−132	−77	143	
15/2	60	189	129	23	−101	−201	−221	−91	273
Matrix elements of $O_{60} = 231J_z^6 - 315J(J+1)J_z^4 + 735J_z^4 + 105J^2(J+1)^2J_z^2 - 525J(J+1)J_z^2 + 294J_z^2 - 5J^3(J+1)^3 + 40J^2(J+1)^2 - 60J(J+1)$									
7/2	1260	−5	9	−5	1				
9/2	5040	−8	6	10	−11	3			
11/2	7560	−20	4	25	11	−31	11		
13/2	2160	−200	−25	185	227	−11	−319	143	

Matrix elements of $O_{66} = \frac{1}{2}(J_+^6 + J_-^6)$

J	F	$(\pm 7/2 \| \mp 5/2)$	$(\pm 9/2 \| \mp 3/2)$	$(\pm 11/2 \| \mp 1/2)$	$(\pm 13/2 \| \pm 1/2)$	$(\pm 15/2 \| \pm 3/2)$
7/2	360	$\sqrt{7}$				
9/2	360	14	$2\sqrt{21}$			
11/2	360	$28\sqrt{3}$	$7\sqrt{30}$	$\sqrt{462}$		
13/2	720	$21\sqrt{10}$	$7\sqrt{66}$	$7\sqrt{33}$	$\sqrt{429}$	
15/2	$360\sqrt{11}$	$42\sqrt{5}$	84	$4\sqrt{273}$	$7\sqrt{39}$	$\sqrt{(35 \times 13)}$

Table 7b (continued)

Matrix elements of $O_{20} = 3J_z^2 - J(J+1)$

J	F	M_J: 0	± 1	± 2	± 3	± 4	± 5	± 6	± 7	± 8
1	1	−2	1							
2	3	−2	−1	2						
3	3	−4	−3	0	5					
4	1	−20	−17	−8	7	28				
5	3	−10	−9	−6	−1	6	15			
6	3	−14	−13	−10	−5	2	11	22		
7	1	−56	−53	−44	−29	−8	19	52	91	
8	3	−24	−23	−20	−15	−8	1	12	25	40

Matrix elements of $O_{40} = 35J_z^4 - 30J(J+1)J_z^2 + 25J_z^2 - 6J(J+1) + 3J^2(J+1)^2$

J	F	M_J: 0	± 1	± 2	± 3	± 4	± 5	± 6	± 7	± 8
2	12	6	−4	1						
3	60	6	1	−7	3					
4	60	18	9	−11	−21	14				
5	420	6	4	−1	−6	−6	6			
6	60	84	64	11	−54	−96	−66	99		
7	12	756	621	251	−249	−704	−869	−429	1001	
8	420	36	31	17	−3	−24	−39	−39	−13	52

Matrix elements of $O_{60} = 231J_z^6 - 315J(J+1)J_z^4 + 735J_z^4 + 105J^2(J+1)^2J_z^2 - 525J(J+1)J_z^2$
$+ 294_z^2 - 5J^3(J+1)^3 + 40J^2(J+1)^2 - 60J(J+1)$

J	F	M_J: 0	± 1	± 2	± 3	± 4	± 5	± 6	± 7	± 8
3	180	−20	15	−6	1					
4	1260	−20	1	22	−17	4				
5	2520	−40	−12	36	29	−48	15			
6	7560	−40	−20	22	43	8	−55	22		
7	3780	−200	−125	50	197	176	−55	−286	143	
8	13860	−120	−85	2	93	128	65	−78	−169	104

Matrix elements of $O_{66} = \frac{1}{2}(J_+^6 + J_-^6)$

J	F	$(3\|-3)$	$(\pm 4\|\mp 2)$	$(\pm 5\|\mp 1)$	$(\pm 6\|0)$	$(\pm 7\|\pm 1)$	$(\pm 8\|\pm 2)$
3	360	1					
4	360	7	$2\sqrt{7}$				
5	360	28	$14\sqrt{3}$	$\sqrt{210}$			
6	360	84	$14\sqrt{30}$	$7\sqrt{66}$	$2\sqrt{231}$		
7	360	210	$42\sqrt{22}$	$28\sqrt{33}$	$2\sqrt{(21\times 143)}$	$\sqrt{(21\times 143)}$	
8	360	462	$42\sqrt{110}$	$12\sqrt{(13\times 77)}$	$14\sqrt{(3\times 143)}$	$7\sqrt{(13\times 55)}$	$2\sqrt{(14\times 143)}$

[a] From [AbBl70].

[b] The numbers in column F are multiplying factors common to all elements in a row.

the same symmetry properties) by the operator equivalent factor α_J, which is defined by

$$\begin{aligned}\sum_i 3z_i^2 - r_i^2 &\equiv \alpha_J\langle r^2\rangle(3J_z^2 - J(J+1)) \\ &= \alpha_J\langle r^2\rangle O_{20}\end{aligned} \tag{3.11a}$$

where

$$\langle r^2\rangle = \int R_{4f}^2 r^2\, d\tau$$

and hence

$$\begin{aligned}&\left(\cdots JM_J\middle|\sum_i A_{20}(3z_i^2 - r_i^2)\middle|\cdots JM'_J\right) \\ &\quad= A_{20}\langle r^2\rangle\alpha_J(\cdots JM_J|3J_z^2 - J(J+1)|\cdots JM'_J) \\ &\quad= A_{20}\langle r^2\rangle\alpha_J(\cdots JM_J|O_{20}|\cdots JM'_J)\end{aligned} \tag{3.12}$$

where $A_{20}\langle r^2\rangle$ is the crystal field parameter; its relation to otherwise defined parameters is given in Tables 6a and 6b. The other terms can be evaluated by the method shown for V_{20}; for brevity we give only the $q = 0$ terms (the others are in Table 7a):

$$\begin{aligned}&\sum_i 35z_i^4 - 30r_i^2z_i^2 + 3r_i^4 \\ &\quad= \langle r^4\rangle\beta_J(35J_z^4 - 30J(J+1)J_z^2 + 25J_z^2 - 6J(J+1) + 3J^2(J+1)^2) \\ &\quad= \beta_J\langle r^4\rangle O_{40}\end{aligned} \tag{3.11b}$$

$$\begin{aligned}&\sum_i 231z_i^6 - 315r_i^2z_i^4 + 105r_i^4z_i^2 - 5r_i^6 \\ &\quad= \langle r^6\rangle\gamma_J(231J_z^6 - 315J(J+1)J_z^4 + 735J_z^4 + 105J^2(J+1)^2J_z^2 \\ &\qquad - 525J(J+1)J_z^2 + 294J_z^2 - 5J^3(J+1)^3 \\ &\qquad + 40J^2(J+1)^2 - 60J(J+1)) \\ &\quad= \gamma_J\langle r^6\rangle O_{60}\end{aligned} \tag{3.11c}$$

The operator equivalent factors α_J, β_J, and γ_J for the ground terms of the RE ions are collected in Table 8. Matrix elements with the same k (but different q) are evaluated with the same operator equivalent factor. The J/M_J dependence of the crystal field matrix elements as obtained with the operators in (3.11) (and that for other polynomials) can be found in many references [Di68, AbBl70], but for convenience we reproduce them for the diagonal elements and O_{66} in Table 7b [Hu64].

The treatment for operator equivalents just outlined more or less follows historical lines. Buckmaster [Bu62, SmTh66], however, pointed out that by a slightly different normalization the results from the tensor operators can be used.

If the definition

$$C_{kq} = \left(\frac{4\pi}{2k+1}\right)^{1/2} Y_{kq} \tag{2.4}$$

TABLE 8

GROUND STATE CONSTANTS USED IN EVALUATING CRYSTAL FIELD MATRIX ELEMENTS IN THE OPERATOR EQUIVALENT FORMALISM[a,b]

Ce^{3+} $4f^1\ ^2F_{5/2}$	Pr^{3+} $4f^2\ ^3H_4$	Nd^{3+} $4f^3\ ^4I_{9/2}$	Pm^{3+} $4f^4\ ^5I_4$	Sm^{3+} $4f^5\ ^6H_{5/2}$	Tb^{3+} $4f^8\ ^7F_6$
$\frac{6}{7}$	$\frac{4}{5}$	$\frac{8}{11}$	$\frac{3}{5}$	$\frac{2}{7}$	$\frac{3}{2}$
$\frac{-2}{5\cdot 7}$	$\frac{-2^2\cdot 13}{3^2\cdot 5^2\cdot 11}$	$\frac{-7}{3^2\cdot 11^2}$	$\frac{2\cdot 7}{3\cdot 5\cdot 11^2}$	$\frac{13}{3^2\cdot 5\cdot 7}$	$\frac{-1}{3^2\cdot 11}$
$\frac{2}{3^2\cdot 5\cdot 7}$	$\frac{-2^2}{3^2\cdot 5\cdot 11^2}$	$\frac{-2^3\cdot 17}{3^3\cdot 11^3\cdot 13}$	$\frac{2^3\cdot 7\cdot 17}{3^3\cdot 5\cdot 11^3\cdot 13}$	$\frac{2\cdot 13}{3^3\cdot 5\cdot 7\cdot 11}$	$\frac{2}{3^3\cdot 5\cdot 11^2}$
0	$\frac{2^4\cdot 17}{3^4\cdot 5\cdot 7\cdot 11^2\cdot 13}$	$\frac{-5\cdot 17\cdot 19}{3^3\cdot 7\cdot 11^3\cdot 13^2}$	$\frac{2^3\cdot 17\cdot 19}{3^3\cdot 7\cdot 11^2\cdot 13^2}$	0	$\frac{-1}{3^4\cdot 7\cdot 11^2\cdot 13}$

Dy^{3+} $4f^9\ ^6F_{15/2}$	Ho^{3+} $4f^{10}\ ^5I_8$	Er^{3+} $4f^{11}\ ^4I_{15/2}$	Tm^{3+} $4f^{12}\ ^3H_6$	Yb^{3+} $4f^{13}\ ^2F_{7/2}$
$\frac{4}{3}$	$\frac{5}{4}$	$\frac{6}{5}$	$\frac{7}{6}$	$\frac{8}{7}$
$\frac{-2}{3^2\cdot 5\cdot 7}$	$\frac{-1}{2\cdot 3^2\cdot 5^2}$	$\frac{2^2}{3^2\cdot 5^2\cdot 7}$	$\frac{1}{3^2\cdot 11}$	$\frac{2}{3^2\cdot 7}$
$\frac{-2^3}{3^3\cdot 5\cdot 7\cdot 11\cdot 13}$	$\frac{-1}{2\cdot 3\cdot 5\cdot 7\cdot 11\cdot 13}$	$\frac{2}{3^2\cdot 5\cdot 7\cdot 11\cdot 13}$	$\frac{2^3}{3^4\cdot 5\cdot 11^2}$	$\frac{-2}{3\cdot 5\cdot 7\cdot 11}$
$\frac{2^2}{3^3\cdot 7\cdot 11^2\cdot 13^2}$	$\frac{-5}{3^3\cdot 7\cdot 11^2\cdot 13^2}$	$\frac{2^3}{3^3\cdot 7\cdot 11^2\cdot 13^2}$	$\frac{-5}{3^4\cdot 7\cdot 11^2\cdot 13}$	$\frac{2^2}{3^3\cdot 7\cdot 11\cdot 13}$

[a] From [ElSt53a].

[b] g_J is the g-factor; α_J, β_J, and γ_J are the operator equivalent factors defined in Eq. (3.11); all in Russell–unders coupling.

TABLE 9

RELATION BETWEEN THE $\tilde{O}_{kq}$ AND O_{kq} OPERATORS[a]

$\tilde{O}_{20} = \frac{1}{2}O_{20}$	$\tilde{O}_{60} = \frac{1}{16}O_{60}$
$\tilde{O}_{22} = (3/8)^{1/2}O_{22}$	$\tilde{O}_{62} = (105/1024)^{1/2}O_{62}$
$\tilde{O}_{40} = \frac{1}{8}O_{40}$	$\tilde{O}_{63} = (105/256)^{1/2}O_{63}$
$\tilde{O}_{42} = (5/32)^{1/2}O_{42}$	$\tilde{O}_{64} = (63/512)^{1/2}O_{64}$
$\tilde{O}_{43} = (35/16)^{1/2}O_{43}$	$\tilde{O}_{66} = (231/1024)^{1/2}O_{66}$
$\tilde{O}_{44} = (35/128)^{1/2}O_{44}$	

[a] From [SmTh66]. (Copyright by The Institute of Physics).

is used, $\tilde{O}_{kq}$ operators are obtained from the tensor operators by replacing x/r, y/r, z/r by J_x, J_y, and J_z, respectively, with proper allowance for the commutation relations. This then yields

$$\tilde{O}_{20} = \tfrac{1}{2}O_{20} = \tfrac{1}{2}(3J_z^2 - J(J+1)) \tag{3.13}$$

A complete set of $\tilde{O}_{kq}$ operators, which are used frequently in the literature, are found in Table 9.

3.3 The Spin Hamiltonian

Because data from optical experiments are often compared with those from electron spin resonance (ESR) and because spin Hamiltonian (or effective spin, or pseudo spin) terminology is sometimes used in interpreting optical experiments, we now discuss what in ESR is called the spin Hamiltonian. Spin resonance often involves situations where transitions are induced between a few levels, that are remote from all the other levels of the system. In such cases it is useful to describe this limited system by an effective spin $\bar{S}$, chosen such that $2\bar{S} + 1$ is the number of levels being considered. ($\bar{S} = \frac{1}{2}$ characterizes a doublet, etc.) In early applications, this formalism was applied to 3d ions, for which the effective spin was often equal to the real spin (hence the nomenclature). The effective spin of RE ions (except those with an S ground state, like Eu^{2+}, Gd^{3+}, Tb^{4+}), however, is rarely the true spin.

The two main terms in the spin Hamiltonian are the Zeeman term (interaction of the magnetic moment of the ion with an external field) and the crystal field interaction (often called the fine structure splitting); with those two terms we obtain†

$$\mathscr{H} = \mu_B(g_x H_x \bar{S}_x + g_y H_y \bar{S}_y + g_z H_z \bar{S}_z) + \sum_{k,q} D_{kq} O_{kq} \qquad (3.14)$$

g_x, g_y, and g_z are usually called g-factors (they are really the principal axes of the g-tensor), although they are not the Landé g_J-factor. They represent merely a method for measuring the splitting in a magnetic field, allowing for anisotropy. In optical experiments the Zeeman splitting is often expressed by the splitting factor $s_j = dW_{jm}/\mu_B dH_j$ ($j = x, y, z$) where W_{jm} is the splitting energy in a magnetic field with respect to the energy $H = 0$; thus for a state characterized by $\bar{S} = \frac{1}{2}$ we have $2s_j = g_j$ ($j = x, y, z$), but for a system with $\bar{S} = 1$ we have $s_j = g_j$. In the frequently encountered case of axial symmetry, $g_z = g_\parallel$ ($s_z = s_\parallel$) and $g_x = g_y = g_\perp$ ($s_x = s_y = s_\perp$).

We shall use the s- and g-factors both because they are used interchangeably in the literature and because some numbers may just sound very unusual if converted to the other definition. If J is replaced by S, the O_{kq} in eq. (3.14) are the same as in Table 7a. The D_{kq} parameters in eq. (3.14) are not necessarily related to any crystal field parameter mentioned so far, because they are used (possibly) to describe only a part of a total system. If, however, the spin Hamiltonian notation is applied to, for example, a crystal field split ground term of an RE ion, we see that $D_{20} = A_{20}\langle r^2\rangle\alpha_J$, etc.

3.4 Examples of Crystal Field Parametrization

We now sketch two examples that show the value of the crystal field parametrization scheme. First, for neodymium ethylsulfate ($Nd(C_2H_5SO_4)_3{\cdot}9H_2O$) we try to deduce, from the material available so far, the crystal field splitting of the $^4I_{9/2}$ ground term. Using spin Hamiltonian notation with the crystal field parameters in Table 10, the operator equivalent factors in Table 8, and the matrix elements in

† In the spin Hamiltonian the crystal field parameters are usually denoted by B_{kq}[AbB170]; in order to avoid confusion with the crystal field parameter B_{kq} we use the "neutral" notation D_{kq}.

TABLE 10

CRYSTAL FIELD PARAMETERS $A_{kq}\langle r^k\rangle$[a]

	$A_{20}\langle r^2\rangle$ (cm^{-1})	$A_{40}\langle r^4\rangle$ (cm^{-1})	$A_{60}\langle r^6\rangle$ (cm^{-1})	$A_{66}\langle r^6\rangle$ (cm^{-1})	Reference
(a) For $RE(C_2H_5SO_4)_3 \cdot 9H_2O$ and RE dissolved in $La(C_2H_5SO_4)_3 \cdot 9H_2O$					
Ce	+9	−42	−45	+680	[Bi67]
Pr	+23	−80	−44	+695	[Hü62b]
Pr(La)	+15	−88	−49	+548	[Gr63a]
Nd	+58	−68	−43	+595	[GrSa63]
Sm	+60	−64	−40	+575	[HiWh66]
Sm	+78	−53			[Hü62b]
Sm(La)	+45	−25	−30	+450	[LaJe66]
Eu	+80	−63	−39	+510	[Ju59]
Gd	+100	−69	−34	+520	[Wy66]
Tb	+110	−75	−34	+465	[Hü62b]
Dy	+124	−79	−31	+492	[PoOr61] [Hü62b]
Dy	+144	−85	−33	+535	[HiWh66]
Ho	+125	−79	−30	+391	[Hü62b]
Ho	+143	−68	−32	+391	[Pe70]
Er	+126	−81	−31	+387	[Er61]
Er	+119	−74	−30	+376	[HiWh66]
Tm(La)	+130	−71	−29	+433	[WoRi61]
Yb	+155	−58	−26	+473	[WeRe68]
(b) For $RECl_3$ and RE dissolved in $LaCl_3$					
Ce	+65	−41	−64	+399	[HeOr65, BaVe69]
Pr	+47	−41	−40	+405	Ma61, HuRa63]
Pr(La)	+50	−40	−39	+397	[Ju57]
Pr(La)	+54	−43	−42	+444	[CaCr77]
Nd	+98	−39	−44	+443	[Ei63b]
Nd(La)	+104	−36	−45	+426	[Ju59]
Nd(La)	+82	−42	−45	+438	[CaCr76]
Pm(La)	+72	−49	−42	+426	[CrCr76]
Sm	+81	−23	−44	+426	[AxDi62]
Sm(La)	+93	−34	−39	+446	[CaCr77]
Eu(La)	+89	−38	−51	+495	[ShDi63]
Eu(La)	+95	−36	−50	+499	[CaCr77]
Gd(La)	+97	−42	−30	+290	[PiDi67]
Gd(La)	+108	−34	−43	+450	[CaCr77]
Tb(La)	+92	−40	−30	+290	[ThSi63]
Tb(La)	+143	−36	−29	+286	[CaCr77]
Dy(La)	+90	−40	−23	+253	[CrDi61]
Dy(La)	+97	−41	−29	+272	[CaCr77]
Ho(La)	+122	−45	−28	+280	[DiPa64]
Ho(La)	+114	−34	−28	+277	[RaKr67, 68]
Ho(La)	+108	−35	−28	+279	[CaCr77]
Er(La)	+94	−37	−27	+265	[Ei63a]
Er(La)	+108	−34	−26	+268	[CaCr77]

[a] From [AbBl70].

Table 7b, we obtain the crystal field matrix elements

$$(\pm\tfrac{9}{2}|V|\pm\tfrac{9}{2}) = +42.5\ \text{cm}^{-1}$$
$$(\pm\tfrac{7}{2}|V|\pm\tfrac{7}{2}) = -139.6\ \text{cm}^{-1}$$
$$(\pm\tfrac{5}{2}|V|\pm\tfrac{5}{2}) = +67.0\ \text{cm}^{-1}$$
$$(\pm\tfrac{3}{2}|V|\pm\tfrac{3}{2}) = +66.2\ \text{cm}^{-1}$$
$$(\pm\tfrac{1}{2}|V|\pm\tfrac{1}{2}) = -36.0\ \text{cm}^{-1}$$

Besides these matrix elements diagonal in J_z, the Hamiltonian also has two nondiagonal (in J_z) nonzero matrix elements:

$$(\pm\tfrac{9}{2}|V|\mp\tfrac{3}{2}) = 71.8\ \text{cm}^{-1}$$
$$(\pm\tfrac{7}{2}|V|\mp\tfrac{5}{2}) = 109.8\ \text{cm}^{-1}$$

This shows that in the crystal field the states with $M_J = \pm\frac{1}{2}$ are not mixed, whereas the other states have a mixing of two $\pm M_J$ values, produced by the nonzero off-diagonal matrix elements. For the evaluation of the energies of the states two 2×2 matrices have to be constructed. The result (if we normalize the energy at the lowest crystal field level) is

$$0\ \text{cm}^{-1}(\pm\tfrac{7}{2}, \mp\tfrac{5}{2});\quad 151.1\ \text{cm}^{-1}(\pm\tfrac{1}{2});\quad 163.7\ \text{cm}^{-1}(\pm\tfrac{9}{2}, \mp\tfrac{3}{2})$$
$$301.6\ \text{cm}^{-1}(\pm\tfrac{7}{2}, \mp\tfrac{5}{2});\quad 319.3\ \text{cm}^{-1}(\pm\tfrac{9}{2}, \mp\tfrac{3}{2})$$

the experimental values are 0 cm^{-1}, 149.4 cm^{-1}, 154 cm^{-1}, 279 cm^{-1}, 311 cm^{-1}, with the same M_J classification, which can be obtained from selection rules and/or Zeeman effect measurements.

The eigenfunctions of the lowest crystal field level are

$$0.915|\pm\tfrac{7}{2}) - 0.395|\mp\tfrac{5}{2})$$

which yields a g-factor for a magnetic field applied parallel to the c axis:

$$g_{\|} = 2(\tfrac{7}{2}\cdot\tfrac{8}{11}\cdot 0.84 - \tfrac{5}{2}\cdot\tfrac{8}{11}\cdot 0.16) = 3.71$$

a result that corresponds well with the experimental value of 3.54 [AbBl70]. This trivial example is meant to show only that the parametrization is indeed useful for cataloging data. In an actual crystal field analysis, of course, we must use the splitting to deduce the crystal field parameters.

The crystal field splitting in many systems has been parametrized in the way just indicated. The procedure adopted in these analyses is as follows: The optical spectra of the system under investigation are taken in order to obtain as many crystal field energy levels as possible. In some cases 30 or more crystal field energy levels have been obtained. These energy levels are then fitted to a Hamiltonian that contains as free parameters ζ, F_2, F_4, and F_6 in order to describe the free ion energy levels, and the crystal field parameters $A_{kq}\langle r^k \rangle$ needed to describe the actual crystal field. Except for cases of very low symmetry, involving a great many crystal field parameters, there are much more experimental data than crystal field parameters and therefore a reliable set of crystal field parameters can be obtained in principle.

In actual calculations, however, a problem can occur. As mentioned, the free ion energy levels are not perfectly described by the Hamiltonian in eq. (2.1). Discrepancies of the order of 50–100 cm^{-1} are possible. If the procedure just described is applied, the crystal field matrix elements nondiagonal in J will immediately try to compensate for this discrepancy, producing an artificial fit. Two approaches have been used to correct this. The better (though quite tedious) method is to incorporate all the other parameters mentioned in Section 2.3 into the fitting routine. This has been done for RE^{3+} in $LaCl_3$ with remarkable results [CrCr76, CaCr76, CaCr77]. However, a more practical and quite successful way can also be used: First, the centers of gravity of the terms are fitted to the free ion Hamiltonian containing only the contributions with ζ, F_2, F_4, and F_6, and the remaining discrepancies are adjusted by small additional energies. With these "perfect" free ion energies the total crystal field Hamiltonian (including J mixing) is then applied, and the small free ion correction energies just mentioned are also varied iteratively. Inclusion of the off-diagonal crystal field matrix elements (in J) is essential in order to obtain a reasonable fit, especially in systems with a large crystal field interaction (e.g., the RE garnets). Another problem lies in the fact that in many instances the description of a measured crystal field splitting by eq. (3.2) is markedly improved if different sets of crystal field parameters are used for different multiplets. This indicates that the charge distribution of the 4f electrons depends on the nature of the state.

An example of such behavior was found, e.g., in the analysis of the crystal field splitting in Sm granets, where different sets of crystal field parameters had to be used for the ^6H and the ^6F multiplet in order to produce satisfactory fits (see Table 12).

This means that one can no longer regard all 4f electrons as equal but has to introduce two and more electron operators into the crystal field Hamiltonian (eq. (3.2)). This, however, introduces a great number of

† For a discussion on how to go beyond the simple crystal field parametrization scheme see also [Ne77a, Ne77b, CoBa78].

additional parameters [BiNe70, Ju77a].† Judd [Ju77b] has, however, found an easy way to handle the leading contribution to the two electron Hamiltonian. He points out that the strong attractive exchange forces between 4f electrons with parallel spin leads to a less extended wavefunction and therefore also to a smaller crystal field parameter. The easiest way to take this into account is to replace the operator $C_{kq}(\vartheta_i, \varphi_i)$ in eq. (3.2) by

$$C_{kq}(\vartheta_i, \varphi_i) + c_k(\mathbf{S}\cdot\mathbf{s}_i)C_{kq}(\vartheta_i, \varphi_i) \tag{3.2a}$$

The parameters c_k must be negative if the contraction of the radial function corresponds to parallel spins. This substitution for the operators changes in eq. (3.7b) the matrix element $(f^N SL\| U_k \| f^N SL')$ to

$$(f^N SL\| U_k \| f^N SL') + c_k[S(S+1)/(2S+1)]^{1/2}(f^N SL\| V_{1k} \| f^N SL) \tag{3.7c}$$

This replacement introduces three new parameters c_k ($k = 2, 4, 6$) but it seems to improve matters considerably where the simple crystal field approach leads to not completely satisfactory results.

The crystal field splitting of the $^7F_1$ and the $^5D_1$ term in $Eu(C_2H_5SO_4)_3 \cdot 9H_2O$ may serve as an example. A $J = 1$ state in the hexagonal ethylsulfate lattice is split into two levels (see Section 3.10) and the splitting is proportional to the operator equivalent factor α_J. Numerically one finds $\alpha(^5D_1)/\alpha(^7F_1) = 0.24$ [Ju77b], whereas the ratio of the crystal field splittings is 0.12 [HeJo57]. Using the substitution in eq. (3.7c) one finds

$$\alpha(^5D_1)/\alpha(^7F_1) = \tfrac{11}{45} + \tfrac{79}{90}c_2$$

A value of $c_2 = -0.14$ brings the measured and calculated ratio of the crystal field splittings of the $^5D_1$ and $^7F_1$ state into agreement. Judd [Ju77b] points out that similar small corrections also improve substantially the crystal field analysis in other cases.

The crystals with point symmetry D_{3h} (and C_{3h}) at the RE site have been the most thoroughly investigated not only by optical spectroscopy but also by ESR and such macroscopic magnetic measurements as specific heat and susceptibility. For the ethylsulfates and chlorides the reasons are that they are still quite symmetric (only four crystal field parameters), are easy to grow (very easy in the case of ethylsulfates), and can be investigated over the whole range of the RE ions, if not in the pure form (e.g., the ethylsulfates), at least as dopants (e.g., in $LaCl_3$). The bromides and hydroxides have been less well investigated because it is difficult to grow them in single crystal form.

Although crystal field theory in its present form was originally tested with ground state properties, the most reliable crystal field analyses have used data obtained by optical spectroscopy. In addition, this technique has the advantage of being applicable to crystals of all concentrations, whereas ESR, for example, is applicable only with dilute specimens. Used to predict or interpret ground state properties, crystal field analyses based on optical data have had remarkable success. Complete agreement, however, can never be achieved because the crystal field Hamiltonian is only an approximate one, and small discrepancies always remain when optical data are fitted to it.

Table 10 presents the crystal field parameters for the ethylsulfate and trichloride lattices obtained by analyzing optical data. It shows quite consistent results. Of the various RE ions, so far only the $PrCl_3$ data have been subjected to a thorough analysis in terms of a priori interpretation from the real crystal potential (see Section 3.5).

The results for the trichlorides by Carnall *et al.* [CaCr77] have been obtained by using 19 terms in the free ion Hamiltonian and should therefore be the most reliable ones, although the deviations from the values obtained by less sophisticated analyses are not too significant.

With respect to the results for Nd^{3+}:$LaCl_3$, the system for which perhaps the most detailed crystal field analysis has been performed, the recent data [CrCr76] were obtained with a 23-parameter fit (although some parameters were constrained) to 101 crystal field energy levels, and the mean deviation between calculated and experimental energy levels is 8 cm^{-1}—quite remarkable, considering that energy levels up to about 40,000 cm^{-1} were used. On the other hand, the original calculation by Judd [Ju59] was performed with much less experimental material (and probably with a desk calculator). In view of this, his results are also quite remarkable (the deviation between his parameters and those of [CaCr76] is less than 10% except for the $A_{20}\langle r^2\rangle$ parameter, where it amounts to 25%; see Table 10). Thus, such seemingly naive calculations [Ju59] reproduce the physics of the problem very well.

To conclude this section, we shall describe results of crystal field analysis in a garnet system [vVl66]. RE garnets have the general formula $3(RE)_2O_3{\cdot}5(ME)_2O_3$ where ME = Al, Ga, or Fe [Pa58]. These crystals have commanded attention because of the interesting magnetic properties of those with ME = Fe (chiefly $3Y_2O_3{\cdot}Fe_2O_3$) and because $3Y_2O_3{\cdot}5Al_2O_3$ (YAG) is an extremely useful laser host material for the 1.06 μm line of Nd^{3+}. We shall deal with Dy^{3+} since $3Dy_2O_3{\cdot}5Al_2O_3$ (dysprosium aluminum garnet, DAG) is one of the classical Ising systems that has been intensively investigated [Wo65], so that the crystal field in this system is of particular interest.

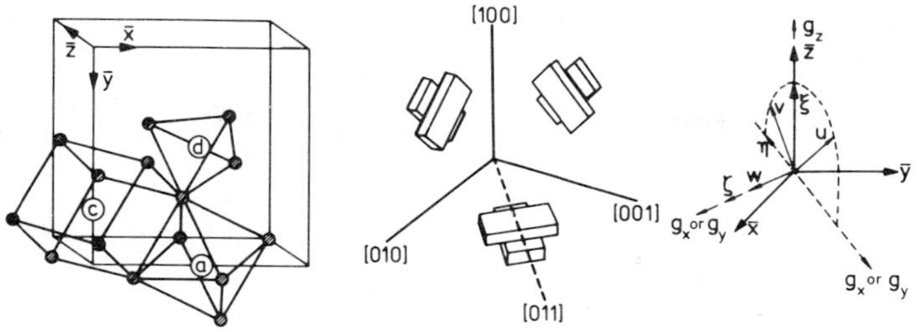

Fig. 10. Crystallographic data and coordinate systems in RE garnets.
Left: part of the crystal structure; the oxygen environments of the octahedral (a) sites, the tetrahedral (d) sites, (occupied by Al, Ga, or Fe ions) and the Re (c) sites with D_2 symmetry are shown. $\bar{x}$, $\bar{y}$, $\bar{z}$ are the unit cell directions. Middle: schematic orientation of the D_2 symmetry axes of the six magnetically inequivalent RE sites [DiWa61]. Right: coordinate system used for the RE ions in the garnet lattice: ($\bar{x}$, $\bar{y}$, $\bar{z}$ coordinate system of the crystallographic unit cell, g_x, g_y, g_z axes of the g-tensor; η, ζ, ξ local axes of the D_2 symmetry; u, v, w axes of the pseudocube).

The overall crystal structure of the garnets is cubic and the position of the various ions is shown in Fig. 10a. The unit cell contains four molecules. The 24 RE ions occupy the c sites, which are surrounded by eight oxygen ions in a near cubic coordination. The ME ions occupy two different sites: 24 are located at the a sites in the middle of an oxygen octahedron and 16 are located at the d sites in the middle of an oxygen tetrahedron [GeGi57]. The exact point symmetry of the RE ions is D_2, and there are six magnetically inequivalent sites (crystallographically equivalent) per unit cell. If the local D_2 symmetry is regarded as an octahedron, the positions of the six octahedra with respect to the coordinate system of the unit cell can be seen in Fig. 10b [DiWa61]. We are therefore dealing with three different coordinate systems, which can be seen in Fig. 10c [HuWo64]:

- $\bar{x}, \bar{y}, \bar{z}$: coordinate system of the cubic crystallographic unit cell.
- ξ, η, ζ: local coordinate system of an RE ion on a D_2 symmetry site.
- g_x, g_y, g_z: principal axis of the g-tensor with respect to which the magnetic properties are defined; the direction of this coordinate system agrees with that of the unit cell axes; however, it is rotated with respect to it. The g_z axis of this system coincides with that of the coordinate system of the unit cell ($\bar{z}$); g_z and g_y, however, coincide with the crystallographic [110] and [1$\bar{1}$0] directions and not with the $\bar{x}$ and $\bar{y}$ axes.

Then there is still the coordinate system used if the crystal symmetry at the site of the RE ion is approximated by a cube (called pseudocube)

u, v, w: axis of the pseudocube of oxygen ions surrounding the RE ion.

We see that the g-tensor axes g_x and g_y (which are indistinguishable) do not agree with the $\bar{x}$ and $\bar{y}$ axes of the crystallographic coordinate system, but point into the [110] and [1$\bar{1}$0] direction, respectively. The g-tensor is usually measured by rotating a crystal in such a way that a magnetic field is swept through a (110) plane; this plane contains a [001] direction yielding g_z for two sites and a [110] direction yielding g_x and g_y for one of each of these two sites, thus making it possible to determine the principal values of the g-tensor. We have to consider the g-factor in some detail here because, due to many crystal field parameters [KaHa73], only a crystal field analysis involving g-factor data is reliable.

The crystal field Hamiltonian for D_2 symmetry has the form (in spin Hamiltonian notation)

$$\begin{aligned}\mathscr{H}_{CF} = {} & \alpha_J(A_{20}\langle r^2\rangle O_{20} + A_{22}\langle r^2\rangle O_{22}) \\ & + \beta_J(A_{40}\langle r^4\rangle O_{40} + A_{42}\langle r^4\rangle O_{42} + A_{44}\langle r^4\rangle O_{44}) \\ & + \gamma_J(A_{60}\langle r^6\rangle O_{60} + A_{62}\langle r^6\rangle O_{62} + A_{64}\langle r^6\rangle O_{64} + A_{66}\langle r^6\rangle O_{66})\end{aligned} \tag{3.15}$$

showing that nine crystal field parameters have to be determined.

The energy level scheme of Dy^{3+} (Fig. 5) shows that the multiplets ^6H and ^6F are well separated from the higher excited terms; hence, they can be treated as a separate system. The crystal field matrix elements between all the ^6H and ^6F terms were included in the calculations [GrHü69a]. The Coulomb and spin–orbit energies, however, were not introduced explicitly; instead, a zero crystal field energy was attributed to each free ion energy level and was used in the fitting procedure. By using crystals with various concentrations of yttrium, practically all the crystal field energy levels of the ^6H and ^6F terms could be obtained. For the fitting procedure, those of $Dy^{3+}_{0.01}$: YAG were used (complemented where necessary by data from $Dy^{3+}_{0.1}$: YAG crystals). A comparison of the calculated and experimental crystal field energy levels is shown in Table 11 and crystal field parameters for various garnet systems are given in Table 12. The garnets are systems with relatively large crystal fields, as compared, e.g., to the ethylsulfates. Therefore, the data in Table 11 give a good idea how well (or badly) the crystal field parametrization works. We note that the $^6F_{11/2}$ and $^6H_{9/2}$ terms and $^6F_{9/2}$ and $^6H_{7/2}$ terms overlap completely,

TABLE 11

COMPARISON OF CALCULATED AND EXPERIMENTAL CRYSTAL FIELD ENERGY LEVELS IN Dy^{3+} : $3Y_2O_3 \cdot 5Al_2O_3$ [a]

Term	E (Calc.) (cm^{-1})	E (Obs.) (cm^{-1})	Term	E (Calc.) (cm^{-1})	E (Obs.) (cm^{-1})
$^6F_{3/2}$	13368	13366	$^6H_{11/2}$	6106	6111
	13357	13359		6099	6102
$^6F_{5/2}$	12662	12674		6084	6087
	12524	12523		6042	6043
	12502	12494		5952	5954
$^6F_{7/2}$	11310	11310		5932	5926
	11252	11265	$^6H_{13/2}$	3945	3951
	11243	11235		3819	3821
	11057	11053		3779	3777
$^6H_{5/2}$	10501	10503		3717	3720
	10362	10354		3695	3698
	10245	10252		3595	3590
$^6F_{9/2}$	9662	9656		3563	3564
$^6H_{7/2}$	9388	9380	$^6H_{15/2}$	732	741
	9371	9367		517	
	9326	9334		474	
	9309	9315		234	233
	9297	9308		172	175
	9088	9075		101	101
	9042	9041		57	59
	8999	9007		0	0
$^6F_{11/2}$	8271	8285			
$^6H_{9/2}$	8202	8218			
	7990	7994			
	7966	7964			
	7949	7942			
	7908	7909			
	7814	7802			
	7768	7760			
	7747	7742			
	7669	7673			
	7640	7641			

[a] From [GrHü69a].

TABLE 12

CRYSTAL FIELD PARAMETERS IN GARNETS

(a) Aluminum garnets

	Nd^{3+} : YAG[a]	Sm^{3+} : YAG[b] $^6H/^6F$[r]	Eu^{3+} : YAG[c]	Tb^{3+} : YAG[d]	TbAG[e]	Dy^{3+} : YAG[f]	DyAG[g]	Er^{3+} : YAG[h]	Yb^{3+} : YAG[i]
$A_{20}\langle r^2\rangle$	−208	−184/68	−191	−217	−168	−178	−183	−156	−12
$A_{22}\langle r^2\rangle$	305	319/667	434	244	391	298	309	267	352
$A_{40}\langle r^2\rangle$	−339	−246/−154	−282	−296	−291	−287	−305	−281	−151
$A_{42}\langle r^4\rangle$	409	−165/345	317	262	220	262	213	221	102
$A_{44}\langle r^4\rangle$	1069	1373/1201	823	984	936	1066	1120	781	692
$A_{66}\langle r^6\rangle$	70	45/33	64	58	44	43	44	32	107
$A_{62}\langle r^6\rangle$	−197	−164/129	−154	−167	−162	−115	−151	−136	−767
$A_{64}\langle r^6\rangle$	1113	1038/638	879	840	753	731	688	649	1009
$A_{66}\langle r^6\rangle$	−41	−41/357	−84	−26	−153	−80	−119	−178	−230

(b) Gallium garnets

	Sm^{3+} : YGG[j] $^6H/^6F$[r]	EuGG[k]	DyGG[l]	Dy^{3+} : YGG[m]	ErGG[n]	Er^{3+} : YGG[o]	Yb^{3+} : YGG[p]	Yb^{3+} : YGG[q]
$A_{20}\langle r^2\rangle$	−52/52	−80	−34	−22	−52	−14	−86	−70
$A_{22}\langle r^2\rangle$	126/−9	237	192	141	72	89	279	288
$A_{40}\langle r^4\rangle$	−231/−189	−263	−274	−282	−222	−238	−193	−177
$A_{42}\langle r^4\rangle$	46/148	200	195	181	246	255	159	221
$A_{44}\langle r^4\rangle$	1283/661	993	985	1040	978	920	535	621
$A_{60}\langle r^6\rangle$	56/22	53	37	36	31	33	72	49
$A_{62}\langle r^6\rangle$	−21/25	−92	−104	−108	−71	58	−229	−462
$A_{64}\langle r^6\rangle$	949/656	839	681	725	657	645	1315	822
$A_{66}\langle r^6\rangle$	88/−233	−176	−93	−57	−55	−70	−233	−233

[a] [MoWo76, ToWe75, KoGe64a, FeTi65, Ne74]. [b] [GrHü69a]. [c] [MoWo76, Ko64]. [d] [MoWo76, Ko64]. [e] [BoGr73]. [f] [MoWo76, GrHü69a]. [g] [WaLe71]. [h] [MoWo76, KoGe64b]. [i] [BuWi67]. [j] [GrHü69b]. [k] [BoGr73]. [l] [WaLe71]. [m] [GrHü69a]. [n] [OrHü69]. [o] [CrMo67]. [p] [HuWo64]. [q] [BuWi67]. [r] $^6H/^6F$ indicates that different sets of crystal field parameters have been used for these two multiplets.

making off-diagonal crystal field matrix elements very large. Still, in these cases too, the approach used reproduces the experimental data quite well.

Of interest in DAG is how well the magnetic properties are described, particularly the large anisotropy of the ground state (lowest crystal field level of the ground term) g-factor, which is the origin of DAG's interesting magnetic properties. Table 13 lists a number of g-factors measured by optical spectroscopy, comparing them with calculated results. These calculations were performed in the local ξ, η, ζ coordinate system, and the results were then transformed into the g_x, g_y, g_z coordinate system for a meaningful comparison with the experimental results. The experimental g-factors are reproduced surprisingly well by the theory. Of course, the calculated g-factors have signs that were not measured experimentally and could have been measured only with circularly polarized light [KaHa72b]. The calculated ground state g-factor shows excellent agreement with the one measured by ESR [WoBa62, BaHu63] (which is much more accurate than optical spectroscopy), yielding

$$(|2s_x| = |g_x| = 0.7; \quad |2s_y| = |g_y| = 0.5; \quad |2s_z| = |g_z| = 17.7)$$

The eigenfunctions of the lowest crystal field level (ground state) of the $^6H_{15/2}$ ground term is (where we abbreviate $^6H_{15/2}$ as $\frac{15}{2}$)

$$\begin{aligned} |\pm)_z = {} & 0.846|\tfrac{15}{2}, \mp\tfrac{15}{2}) - 0.497|\tfrac{15}{2}, \mp\tfrac{11}{2}) \\ & + 0.100|\tfrac{15}{2}, \mp\tfrac{7}{2}) + 0.149|\tfrac{15}{2}, \mp\tfrac{3}{2}) \\ & - 0.000|\tfrac{15}{2}, \pm\tfrac{1}{2}) - 0.020|\tfrac{15}{2}, \pm\tfrac{5}{2}) \\ & + 0.041|\tfrac{15}{2}, \pm\tfrac{9}{2}) - 0.019|\tfrac{15}{2}, \pm\tfrac{13}{2}). \end{aligned}$$

TABLE 13

CALCULATED AND OBSERVED g-FACTORS IN Dy^{3+} : YAG[a]

Theoretical				Experimental			
W (cm^{-1})	g_x	g_y	g_z	W (cm^{-1})	$\lvert g_x\rvert$	$\lvert g_y\rvert$	$\lvert g_z\rvert$
0	−0.8	−0.5	17.8	0	1.6	1.6	18.0
57	−4.4	3.4	−11.3	59	4.3	4.3	11.9
101	4.9	−1.4	10.4	101	5.7	5.7	10.4
172	−3.6	6.2	−7.1	175	3.2	3.2	8.0
3563	0.8	−0.3	−14.9	3564	1.1	1.1	14.3
3595	−0.6	0.3	13.2	3590	0.6	0.6	13.1

[a] From [GrHü69a].

Here we see clearly that the ground state is a predominantly $|\frac{15}{2}, \mp\frac{15}{2}\rangle$ state, which explains the magnitude of the g_z-factor (18 as compared with 20 for a pure $|\pm\frac{15}{2}\rangle$ state) and the pronounced anisotropy of the g-factor.

3.5 Model Description of the Crystal Field

In this section we consider the theoretical consequences of experimental results on the crystal field splitting of RE ions presented so far. We have seen that the Stevens [Ste52] parametrization scheme is very successful in describing the experimental data. As a matter of fact, considering the approximations in the various fitting procedures, the results are as good as could possibly be expected. This leads to two questions.

(1) Why is the parametrization scheme used for the analysis of the crystal field interaction so successful?
(2) Is it possible to determine the magnitude of the crystal field parameters from an a priori calculation?

The first question was basically answered earlier when we introduced the crystal field parameters. The parametrization scheme makes no specific assumptions about the nature of the crystal field (as does, e.g., the point charge model) and therefore holds equally well for more or less ionic crystals. The two main restrictions entering the scheme are that we assume that the crystal field acts on all 4f electrons independently and equally, and that the radial distribution function of the 4f electrons is the same for all terms considered. Both these approximations seem justifiable for 4f electrons and therefore make the parametrization scheme's success on a qualitative basis understandable.

Newman [Ne71, BrNe67] has tried to put the parametrization scheme on a more mathematical basis in what he calls the superposition model. The reasoning underlying this model can be found in earlier work [HuWo64, JoPa63, SchäJo65]. The model assumes that each ion in the crystal contributes separately and independently to the crystal field of a reference ion. Therefore, all like ions contribute the same amount if their distance is the same and if the geometrical factors are taken into account. We can therefore define intrinsic parameters $\bar{A}_k(R)$, which depend only on the distance R. The experimental parameters can then be written in the form

$$A_{kq}\langle r^k\rangle = \sum_j g_{kq}(j)\bar{A}_k(R_j) \tag{3.16}$$

where the g_{kq} are the geometrical factors, which can be determined from the x-ray structure data, and the $\bar{A}_k(R_j)$ are called intrinsic parameters; j is the running index for the various ions. To make the superposition model applicable we have to introduce the further assumption that only the ligands contribute significantly to the crystal field. This is valid for the $k = 4$ and $k = 6$ terms, whose contributions fall off very rapidly with distance; it is, however, a bad approximation for the $k = 2$ terms, which are therefore not considered in this model.

Analysis of a set of crystal field parameters in terms of the superposition model has the advantage, especially in the case of low symmetry of reducing the number of parameters; it also shows, at least qualitatively, the effect of two- and more-body interactions on the size of the crystal field splitting. It also has value in the a priori calculation of crystal field parameters, which can now be performed in terms of interactions between two ions.

We shall briefly discuss the usefulness of the superposition model with reference to the crystal field analysis in garnet hosts [NeSte69, NeSte71]. A crystal field analysis for a typical garnet system was given in the preceding section. The point symmetry at the RE site is D_2 and nine crystal field parameters are needed to parametrize the observed splitting; if we neglect the $k = 2$ terms of the crystal field development because of their strong dependence on the point charge contribution (which gets large contributions from ions beyond the nearest neighbors, which by definition are excluded from the superposition model), seven experimental parameters still have to be described. The ligands surrounding the RE ions in garnet crystals are eight oxygen ions at two different distances. Therefore, we have to fit four intrinsic parameters to seven experimental ones. (Four parameters are needed because, for the $k = 4$ and $k = 6$ contributions, ligands with two different distances contribute.)

Superposition analysis was performed for the crystal field parameters obtained in erbium gallium garnet (ErGaG) and those obtained for erbium in yttrium gallium garnet (Er^{3+}:YGaG). The results for the simple superposition model analysis appear in the "no distortion" column in Table 14. The values for t_4 and t_6, the power law exponents for the $k = 4$ and $k = 6$ intrinsic parameters, defined by $\bar{A}_k(R_2) = \bar{A}_k(R_1)(R_1/R_2)^{t_k}$ (where R_1 and R_2 are the distances of the ligands 1 and 2, respectively), were not obtained independently in the fitting procedure but were calculated from the intrinsic parameters obtained for R_1 and R_2. An important point is that the crystallographic data are available only for YGaG; the same values were used for ErGaG, which can introduce an uncertainty. From Table 14 we see that the no distortion calculations reproduce the

TABLE 14

CRYSTAL FIELD PARAMETERS AND INTRINSIC PARAMETERS[a]

	ErGaG			Er^{3+} : YGaG		
	Experiment (cm^{-1})	No distortion (cm^{-1})	Angular distortion (cm^{-1})	Experiment (cm^{-1})	No distortion (cm^{-1})	Angular distortion (cm^{-1})
$A_{40}\langle r^4\rangle$	−221	−240	−226	−238	−246	−238
$A_{42}\langle r^4\rangle$	246	243	339	255	254	255
$A_{44}\langle r^4\rangle$	978	839	948	920	860	916
$\bar{A}_4(R_1)$		94.0	88.8		96.6	95.6
$\bar{A}_4(R_2)$		57.6	58.2		58.9	55.4
t_4		13.0	11.3		13.2	14.5
$A_{60}\langle r^6\rangle$	31.5	36	38	33.0	36	33
$A_{62}\langle r^6\rangle$	−71	−41	−77	−58	−24	−58
$A_{64}\langle r^6\rangle$	657	640	625	645	636	642
$A_{66}\langle r^6\rangle$	−55	−26	−69	−70	−31	−70
$\bar{A}_6(R_1)$		23.7	25.7		22.8	23.1
$\bar{A}_6(R_2)$		21.7	16.0		22.3	18.1
t_6		2.4	12.6		0.5	6.6
$\Delta\theta$			−2.3°			−0.2°
$\Delta\phi$			9.1°			9.1°

[a] From [NeSte69]. R_1 and R_2 are the distances of the two different ligands.

"measured" crystal field parameters reasonably well. Not unexpectedly, the quality of the analysis is somewhat better in Er^{3+} : YGaG than in ErGaG. The most evident problem with the analysis is the small value of t_6, especially in light of the large value obtained for t_4. To improve the quality of the fit for Er^{3+} : YGaG it might be argued that introducing Er into the YGaG lattice distorts the immediate environment slightly, so that the YGaG crystallographic parameters are not the correct ones. In principle, we would have to assume that each set of oxygen ions could be in a distorted position; this would introduce six new parameters and make the problem intractable. It was therefore further assumed that the four nearest (nn) oxygen ions do not change their position at all, and that the four next nearest oxygens are on only angularly distorted positions (the distance is not changed). This gives two new parameters, $\Delta\theta$ and $\Delta\phi$ (for an exact definition of these parameters see [NeSte69]), which allow for an angular distortion of the oxygen ions. Altogether, then six parameters are to be determined from seven experimental values. We see in Table 14 that the crystal field parameters calculated by allowing for an angular distortion are quite close to the experimental ones; much more so, of course, in Er^{3+} : YGaG, presumably because we started with the proper positions in the first place. In ErGaG the distortion model is less justified, and can be regarded as an approximate correction for not using the proper x-ray data. The analysis that allows for the angular distortion of the position of the nn oxygen ions markedly increases the value for t_6; it is, however, still somewhat disquieting that it is smaller than t_4 in Er^{3+} : YGaG and only about as large as t_4 in ErGaG, since we might have expected it to be larger. Perhaps we should also keep in mind that the number of parameters to be determined and the experimental data available are very similar, which introduces some uncertainty.

Superposition analysis seems a very useful way to check the consistency of a set of experimentally determined crystal field parameters; it is also a good starting point for an a priori calculation of crystal field parameters. First principle calculations of crystal field parameters that give satisfying results are still rare, having been performed so far only by Newman and his collaborators [Ne71] (for earlier attempts see, e.g., [AxBu66, WaFr67]).

In the early days a priori calculations were performed with the point charge approximation; the results were mostly too large for the $k = 2$ terms, of the right order of magnitude for the $k = 4$ terms, and an order of magnitude too small for the $k = 6$ terms. The point charge calculations were then refined to take into account both the polarizabilities of the ions and the shielding of the lattice electrostatic potential by the closed $5s^2 5p^6$ shells [HuRa63, RaRa67]. The latter mechanism proved especially significant for the $k = 2$ term, thus removing some of the discrepancy in the point charge calculations; for the $k = 4$ and $k = 6$ terms, however, it yielded only slight improvement. Axe and Burns [AxBu66] made the first quantitative attempt to go beyond the point charge model with the molecular orbital [Ba62] treatment of the crystal field splitting for the Tm^{2+}:CaF_2 system. This system, having only one 4f hole and cubic symmetry, is especially easy to handle. Although they had to invoke a number of assumptions, their calculations showed that molecular orbital-type calculations can indeed produce the right order of magnitude for the $k = 4$ and $k = 6$ terms of the crystal field interaction.

The superposition model's success shows that the $k = 4$ and $k = 6$ terms of the crystal field are dominated by the interaction of the central ion with its ligands; in addition, two- and three-body interactions seem small. These findings provide the starting point for the ab initio calculation of the crystal field parameters as performed by Newman and his collaborators for some systems; we shall deal here with the results for $PrCl_3$

because they seem to be the most detailed [ElNe69, ElNe67, BiNe68, CuNe70, BiNe70, Ne71].† In these calculations the crystal is split into two parts: a $PrCl_9^{-6}$ cluster, which is the central ion with all its ligands, and the rest of the crystal; the cluster is then treated "exactly," whereas only the point charge contributions from the rest of the crystal are used. To present the details of the calculations would require a full description of molecular orbital theory. Instead, we shall only discuss briefly the nature of the two most important contributions to the crystal field interaction, namely, overlap and covalency, and then simply present the results of the crystal field parameter calculations [WaFr67, WaFr64c]. For the $PrCl_9^{-6}$ cluster calculation, the superposition model is used: That is, only one Pr–Cl pair is considered. The interaction of the 4f ϕ orbitals and ligand χ orbitals for this pair is calculated. The overlap contribution is more or less a consequence of the nonorthogonality of the ϕ and χ orbitals of like spin. Take as a simple example one 4f orbital $\phi_\uparrow$ and two ligand orbitals $\chi_\uparrow$ and $\chi_\downarrow$; then $\phi_\uparrow$ and $\chi_\downarrow$ are nonorthogonal and the Schmitt orthogonalization produces the three orbitals

$$(1 - S^2)^{-1/2}(\phi_\uparrow - S\chi_\downarrow); \quad \chi_\uparrow; \quad \chi_\downarrow \tag{3.17}$$

where S is the overlap integral. If $\mathscr{H}$ is the one-electron Hamiltonian of the system, the energy shift by the overlap is

$$\Delta W_{\text{ov}} = -2S(\phi_\uparrow|\mathscr{H}|\chi_\uparrow) + S^2((\phi_\uparrow|\mathscr{H}|\chi_\uparrow) + (\chi_\uparrow|\mathscr{H}|\chi_\uparrow)) \tag{3.18}$$

In contrast to the overlap contribution, the covalency involves charge transfer from one center to the other, in our case from the ligand to the central ion. In our model system, since $\phi_\uparrow$ is occupied, a $\chi_\downarrow \rightarrow \phi_\downarrow$ transfer occurs, which gives rise to the new set of orbitals

$$(1 - S^2)^{-1/2}(\phi_\uparrow - S\chi_\uparrow); \quad \chi_\uparrow; \quad (1 + 2S\gamma + \gamma^2)^{-1/2}(\chi_\downarrow + \gamma\phi_\downarrow) \tag{3.19}$$

where γ is the so-called mixing factor. This mixing gives a contribution to the energy of the 4f orbital of the form

$$\Delta W_{\text{cov}} = -2\gamma(\phi_\downarrow|\mathscr{H}|\chi_\downarrow) + (2\gamma S + \gamma^2)(\chi_\downarrow|\mathscr{H}|\chi_\downarrow) - \gamma^2(\phi_\downarrow|\mathscr{H}|\phi_\downarrow) \tag{3.20}$$

These simple considerations are meant only to explain the terms covalency and overlap; they certainly cannot explain the very involved calculations necessary to obtain actual values.

Altogether so far ten different contributions to the crystal field param-

† For a systematic interpretation of crystal field parameters of RE^{3+} : $LaCl_3$ in terms of molecular orbital theory see [Ur78].

TABLE 15

RESULTS OF A FIRST PRINCIPLE CALCULATION OF THE CRYSTAL FIELD PARAMETERS IN $PrCl_3$[a]

Mechanism	$A_{20}\langle r^2\rangle$	$A_{40}\langle r^4\rangle$	$A_{60}\langle r^6\rangle$	$A_{56}\langle r^6\rangle$	Reference
(1) Ligand point charges	158	−17	−2.7	33	[ElNe67]
(2) Remaining point charges	638	8	−0.9	24	[HuRa63]
(3) Dipolar polarization	469	3	−0.3	5	[HuRa63]
(4) Quadrupolar polarization	−705	−14	1.0	−5	[HuRa63]
(5) Charge penetration	−39	17	6.8	−82	[RaRa67][ElNe69]
(6) Overlap	13	−27	−27	326	[ElNe67][ElNe69]
(7) Covalency	7	−14	−13.1	160	[ElNe69]
(8) Ligand–lanthanide exchange charge	−105	8	0.6	−7	[BiNe68]
(9) Ligand–ligand exchange charge	−57	−3	−0.5	6	[CuNe69]
(10) Triangular path contributions	2	−5	−5.1	76	[CuNe69]
Total theory	(381)	−44	−41.2	536	
Experiment	47	−40.6	−39.6	405	[Ma61]

[a] From [Ne71].

eters in $PrCl_3$ have been calculated. Table 15 lists the results of the crystal field parameters calculations and compares them with the experimental parameters. The contributions (1)–(4) are self explanatory. The polarization terms (3), (4) were calculated by using free ion polarizabilities. The charge penetration contribution (5) is a correction to (1) where the effect of the ligands is modified due to the finite size of their charge distribution. The overlap and covalency term have just been explained. The ligand lanthanide exchange charge term (8) comes from the penetration of the ligand charges into the $5s^2 5p^6$ shell; this charge distribution is thereby modified by a contact interaction. A similar contribution occurs by the mutual interpenetration of the ligand charges (9). Finally, there is a three-body interaction between the 4f ion and two ligands, including an overlap integral between two ligands (10). Contributions (9) and (10) are not encompassed by the superposition model but they are small.

The comparison between theory and experiment is quite gratifying for the $A_{40}\langle r^4\rangle$, $A_{60}\langle r^6\rangle$, and $A_{66}\langle r^6\rangle$ terms. We also note that covalency and overlap contribute significantly to the size of the parameters. As expected, the calculated $A_{20}\langle r^2\rangle$ parameter is considerably larger than the experimental one. Also, in contrast to the fourth and sixth order parameters, the second order parameter is given mostly by the point charge and polarization contributions. This fact introduces a number of problems. The polarization contributions were calculated by using free ion polarizations; it is not known how much these values are changed in going to the solid. Moreover, it is known that the electrostatic second order contributions are considerably shielded [WaFr64b, FrWa65]. This has often been expressed as

$$A_{kq}\langle r^k\rangle = (1 - \sigma_k)A_{kq}\langle r^k\rangle \text{ (point charge)} \tag{3.21}$$

In the simplest interpretation the point charge contribution is understood to comprise only the first two terms in Table 15; the next step is to include contributions (3) and (4) as well.

At this point, however, it is not clear how the contributions in Table 15 are affected by shielding. Calculations give values for σ_2 of the order of 0.8 and for σ_4 and σ_6 of the order of 0.1 [SeAr70]. This shows that a reliable theoretical estimate of $A_{20}\langle r^2\rangle$ requires a careful shielding calculation.

3.6 Classification of Crystal Field States and Selection Rules: Survey

Of considerable importance in all spectroscopic techniques are the selection rules that hold for transitions between two energy levels. Selection rules enable us to draw conclusions about the nature of the energy levels between which a transition takes place. In optical spectroscopy, these selection rules have greatly enhanced our understanding of the structure of the energy levels in free atoms and in crystals. The selection rules in crystal spectra are intimately connected with the symmetry of the crystal under consideration. To arrive at an analytical description of this relationship, we therefore need a mathematical concept capable of dealing with the symmetry properties of crystals. This concept is group theory. Since certain symmetry principles are involved in almost all physical theories, group theory plays an important role in many branches of physics, and there are many excellent reviews dealing with the application of this discipline to physics. Nevertheless, students often lack a working knowledge of group theory, so we shall outline those aspects of the theory that are necessary to derive the selection rules.

The situation for RE crystal spectra is somewhat complicated because the literature records two approaches to classifying the energy levels in a crystal. The original one, due to Bethe [Be29], is based on group theory. Later, Hellwege [He49] (for a brief introduction see [He76]) developed a method of classifying crystal field levels that started with the zero order wavefunctions in the crystal and ended with labeling the states by so-called crystal quantum numbers. This approach can be very useful for interpreting experimental results, although in some cases, due to ambiguities, two crystal quantum numbers are used to characterize one energy level. Because group theory is so widely used we give only the group theoretical derivation of the classification of crystal field states and then give, without proof, the classification by crystal quantum numbers (see [He49, He76]) in such a way that the connection with the group theoretical results can be made.

Though we do not intend to review group theory, we want to outline those elements that are needed in order to apply the theory to crystal spectroscopy. For details, see the books by Tinkham [Ti64] and Lax [La74], as well as Wybourne's book [Wy65] and the article by Fick and Joos [FiJo57].

3.7 Sketch of Group Theory

Definition and Example

A group is defined as a set, or group, of elements $A, B, C, \ldots$, where a multiplication of two elements obeys the following rules:

(1) The product of any two elements in the group is an element of the group.

(2) For any three elements in the group, $A(BC) = (AB)C$, the associative law holds.

(3) The group contains a unit element E, such that $(EA) = (AE) = A$.

(4) Each element in the group has an inverse, also a member of the group, such that $(AA^{-1}) = (A^{-1}A) = E$.

To make these definitions more real, we present an example. In what follows, we shall identify the group elements with symmetry operations, a convenient way to describe crystal symmetry. These symmetry operations are

E: the identity operation.

C_n: rotation about an axis of symmetry $2\pi/n$.

σ_h: reflection in a plane of symmetry perpendicular to the principal axis of rotation.

σ_v: reflection in a plane of symmetry that contains the principal axis of rotation.

i: inversion through a center of symmetry.

S_n: rotation about a symmetry axis by $2\pi/n$ followed by a reflection in a plane perpendicular to the axis of rotation.

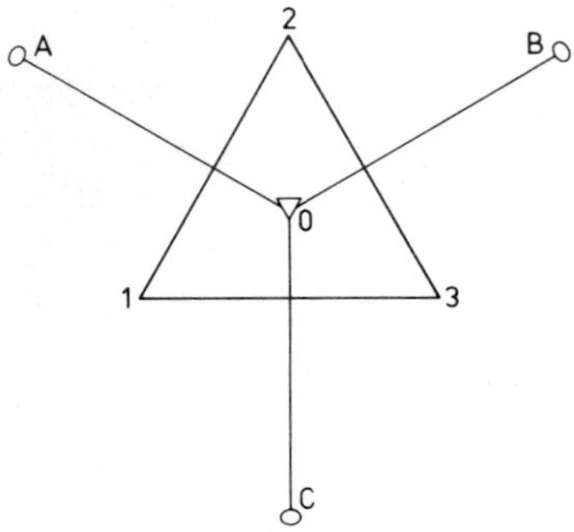

Fig. 11. The symmetry elements of a unilateral triangle (D_3 symmetry); there are a threefold axis (Δ) and three twofold axes (O) [Ti64, Wy65].

The sum of these symmetry operations constitute a group that transforms a symmetrical figure into itself. An equilateral triangle (Fig. 11) [Ti64, Wy65], for example, has these three classes of symmetry operations:

E: the identity operation.
C_3, C_3': clockwise or counterclockwise rotation by $2\pi/3$ about the principal axis.
$C_2(1), C_2(2), C_2(3)$: rotation by π around one of the twofold axes $0A$, $0B$, $0C$

The group definitions enable us to construct the multiplication table of the group. For crystals this group is called D_3, and its multiplication table is Table 16, which shows that each successive application of any two group operations is equivalent to the application of some single group operation.

TABLE 16

MULTIPLICATION TABLE OF THE D_3 GROUP[a]

	E	C_3	C_3'	$C_2(1)$	$C_2(2)$	$C_2(3)$
E	E	C_3	C_3'	$C_2(1)$	$C_2(2)$	$C_2(3)$
C_3	C_3	C_3'	E	$C_2(3)$	$C_2(1)$	$C_2(2)$
C_3'	C_3'	E	C_3	$C_2(2)$	$C_2(3)$	$C_2(1)$
$C_2(1)$	$C_2(1)$	$C_2(2)$	$C_2(3)$	E	C_3	C_3'
$C_2(2)$	$C_2(2)$	$C_2(3)$	$C_2(1)$	C_3'	E	C_3
$C_2(3)$	$C_2(3)$	$C_2(1)$	$C_2(2)$	C_3	C_3'	E

[a] From [Wy65].

Class, Representation, Reducibility, Character

Two group elements A and B are called conjugate if the relation

$$B = C^{-1}AC$$

holds, where C is any member of the group. All group elements that are conjugate to each other are said to belong to a class. The meaning of this assertion can again be demonstrated with the equilateral triangle in Fig. 11, in which the relation

$$C_2(1) = C_3^{-1}C_2(2)C_3$$

holds. (C_3 rotates 1 into 2, etc; $C_2(2)$ rotates 2 into 3; $C_3^{-1} = C_3'$ rotates 3 into 2; the same effect can be obtained by performing the $C_2(1)$ operation, which rotates 1 into 2.) By this procedure we can see that the two C_3 operations and the three C_2 operations each belong to one class; in addition, the identity operation forms a class of its own. Physically speaking, a class always contains those symmetry elements which have the same physical property.

The property of having different classes has some relation to what is called the representation of a group; a group has as many (irreducible) representations as it has classes. The representations of a group can be expressed as quadratic matrices M_i. For each group element a matrix is constructed in such a way that the multiplication table remains valid:

Group element		Matrix		
A	$\longrightarrow$	M_A	results in $\longrightarrow$	if $A \cdot B = C$ is valid, then $M_A \cdot M_B = M_C$
B	$\longrightarrow$	M_B		
C	$\longrightarrow$	M_C		

Again resorting to the equilateral triangle as an example, we find the representation table given in Table 17 for the group D_3.

In the following it will be assumed that the representations consist of unitary matrices, which can always be made by a transformation. The order of the matrices of a representation is called the dimension of the representation; thus in our example we have one two-dimensional and two one-dimensional representations.

TABLE 17

REPRESENTATION TABLE OF THE D_3 GROUP[a]

	E	C_3	C_3'	$C_2(1)$	$C_2(2)$	$C_2(3)$
$^1\Gamma_1$	1	1	1	1	1	1
$^1\Gamma_2$	1	1	1	−1	−1	−1
$^2\Gamma_3$	$\begin{pmatrix} 1 & 0 \\ 0 & 1 \end{pmatrix}$	$\begin{pmatrix} -1/2 & -\sqrt{3}/2 \\ \sqrt{3}/2 & -1/2 \end{pmatrix}$	$\begin{pmatrix} -1/2 & \sqrt{3}/2 \\ -\sqrt{3}/2 & -1/2 \end{pmatrix}$	$\begin{pmatrix} -1 & 0 \\ 0 & 1 \end{pmatrix}$	$\begin{pmatrix} 1/2 & \sqrt{3}/2 \\ \sqrt{3}/2 & -1/2 \end{pmatrix}$	$\begin{pmatrix} 1/2 & -\sqrt{3}/2 \\ -\sqrt{3}/2 & -1/2 \end{pmatrix}$

[a] From [Wy65].

If the representations of a group element A can be written in the form

$$\Gamma(A) = \begin{pmatrix} \Gamma_1(A) & 0 & 0 \\ 0 & \Gamma_2(A) & 0 \\ 0 & 0 & \Gamma_3(A) \end{pmatrix} \tag{3.22}$$

then we say that $\Gamma_1(A)$, $\Gamma_2(A)$ and $\Gamma_3(A)$ are the irreducible representations of the group of which A is an element, and this is generally written as

$$\Gamma(A) = \Gamma_1(A) + \Gamma_2(A) + \Gamma_3(A) \tag{3.23}$$

In the example

$$\Gamma(D_3) = {}^1\Gamma_1 + {}^1\Gamma_2 + {}^2\Gamma_3$$

The traces of the matrices of a representation are called its characters and are denoted by χ; they are used to construct the character tables of a representation. In our example, this table takes the form of Table 18 (for D_3 symmetry).

TABLE 18

CHARACTER TABLE OF THE D_3 GROUP[a]

Representation \ Class	E	$2C_3$	$3C_2$
${}^1\Gamma_1$	1	1	1
${}^1\Gamma_2$	1	1	-1
${}^2\Gamma_3$	2	-1	0

[a] From [Wy65].

The character of a representation is invariant under any similarity transformation. This means that the character of all the matrices belonging to one class are equivalent, and therefore it is not necessary to distinguish different classes; one just notes before the symmetry element characterizing a class the number of times it occurs (see Table 18, row one).

We shall soon see that it is particularly important in crystal field theory to find the number of times p_s that an irreducible representation Γ_{sr} (with character χ_{sr}) is contained in a reducible representation Γ_r (with character χ_r); this number is given by

$$p_s = \frac{1}{n} \sum_{r=1}^{c} h_r \chi_r \chi_{sr}^* \tag{3.24}$$

where n is the order of the group, which is given by the number of distinct elements (number of symmetry operations) in the group; c is the number of classes, or number of irreducible representations Γ_{sr}, into which Γ_r can be decomposed; h_r is the number of elements that the class r contains (number of times that a particular symmetry element occurs).

3.8 Kramer's Degeneracy

For the analysis of the crystal field splitting of RE ions a theorem by Kramers is of considerable practical importance [Kr30]. It states that two crystal field states that can be transformed into each other by reversing the time coordinate (reversing all velocities) have the same energy (Kramers degeneracy). The best known consequence of this theorem is that for ions with an odd number of electrons all crystal field levels are at least doubly degenerate, or that nondegenerate levels in a crystal field can be found only for ions with an even number of electrons. (Doubly degenerate levels are, of course, also possible for ions with an even number of electrons.)

The proof of this theorem is rather long and will be omitted. We give only some arguments that show that the theorem is reasonable. Let us consider a state with $J = \frac{1}{2}$ and one with $J = 1$, in a homogeneous electric field. Since an atom has no permanent electric moment, the electric field can only interact with a moment induced in its electron distribution. Hence, the electric field interacts with the average charge distribution of the atom. The charge distribution is, however, independent of the direction of motion of the electrons (it is also independent of the direction of the spin of the electron, since it does not interact at all directly with the spin) and therefore the interaction energy must be independent of the sign of the directional quantum number M_J. In terms of the states $J = \frac{1}{2}$ and $J = 1$, this means that the electric field cannot split the $J = \frac{1}{2}$ state any further, whereas the $J = 1$ state will be split into an $M_J = 0$ and an $M_J = \pm 1$ state. In odd electron systems (with noninteger values of J) we find, because it is not possible to split a $J = \frac{1}{2}$ state any further by an electric field, that for states with higher angular momentum the $\pm M_J$ degeneracy will also always remain. This conclusion cannot be altered if the electric field has higher multipole moments, because the symmetry of the problem remains. (Of course, there can be fourfold degeneracies if the symmetry of the crystal is high enough.) Similarly, for higher integer angular momentum systems in fields with higher multipole moments there can be at least one singly degenerate state. If, however, there are other states that belong to the same representation, they will also be singlets. Thus if the symmetry of the crystal field is low enough, for a system with integer J, all states *can* be singlets. For high enough crystal symmetry, however, there must not be singlets in an even electron system.

If the Hamiltonian for the crystal field system is

$$\mathscr{H} = \mathscr{H}(\mathbf{r}_i, \mathbf{l}_i, \mathbf{s}_i) \tag{3.25a}$$

then the time reversed operator is

$$\bar{\mathscr{H}} = \mathscr{H}(\mathbf{r}_i, -\mathbf{l}_i, -\mathbf{s}_i) \tag{3.25b}$$

Kramers' theorem now states that if

$$\Psi = \sum a_{JM_J} |JM_J) \tag{3.26a}$$

yields

$$\mathscr{H}\Psi = W\Psi$$

then there exists a Kramers state $\overline{\Psi}$ that has the same energy:

$$\bar{\mathscr{H}}\overline{\Psi} = W\overline{\Psi}$$

Here we obtained $\overline{\Psi}$ from Ψ by the procedure [AbBl70]

$$\overline{\Psi} = \sum_{JM_J} a^*_{JM_J}(-1)^{J-M_J} |J - M_J) \tag{3.26b}$$

The theorem of Kramers somewhat complicates the application of group theory to the classification of crystal field states in odd electron systems, because it requires that we resort to the concept of double groups. Conceptually, this presents nothing new. In what follows results for even and odd electron systems will be given; the exact derivation may be found either in Bethe's [Be29] original work or in other volumes mentioned.

3.9 Group Theory and Crystal Field Energy Levels

So far, we have treated group theory as an abstract mathematical tool. We now show how it can be used in actual calculations [AbBl70, Ti64, FiJo57].

The symmetry of the space of a system will be defined by its symmetry group G with the transformations R. If the Hamiltonian of the system is invariant under all transformations R of the form

$$x' = R \cdot x$$

then if W is an eigenvalue with eigenfunction $\Psi(x)$, $\Psi(Rx)$ is also an eigenfunction of W. Now we can define symmetry operations P_R such that the function $P_R\Psi$ at the position $x' = Rx$ has the same value as the function Ψ at position x:

$$P_R\Psi(Rx) = P_R\Psi(x') = \Psi(x) \qquad \text{or} \qquad P_R\Psi(x) = \Psi(R^{-1}x) \quad (3.27)$$

The transformations P_R form a group G_n that is isomorphous with G. By definition, the points x and $x' = Rx$ are equivalent, and the functions Ψ and $P_R\Psi$ have to be equivalent because they have the same eigenvalue

$$\mathscr{H}\Psi = W\Psi \qquad \mathscr{H}P_R\Psi = WP_R\Psi$$

and therefore

$$\mathscr{H}P_R = P_R\mathscr{H}$$

meaning that $\mathscr{H}$ and P_R commute.† If the eigenvalue W is n-fold degenerate, then there are n linear independent eigenfunctions $\Psi_1, \Psi_2, \ldots, \Psi_n$; the eigenfunctions $P_R\Psi_1 \cdots P_R\Psi_n$ also belong to W and must be linear combinations of the original eigenfunctions.

$$P_R\Psi_i(x) = \sum_{j=1}^{n} D_{ji}(R)\Psi_j(x), \qquad i = 1, \ldots, n \quad (3.28)$$

The D_{ji} (with dimension n) are elements of a matrix D that forms a representation of G; the dimension of the representation is n, the number of linear independent functions Ψ_i. We can thus say that to each n-fold degenerate level there belongs an n-dimensional representation.

For the sake of further discussion we will assume that the original group G is the group of three-dimensional rotations R_3 (this, unlike all the other groups we shall deal with, is a group with an infinite number of elements). Energy levels are characterized by the total angular momentum J; the degeneracy of every energy level will be $n = 2J + 1$. Under the operation of R_3 the eigenfunctions of the free ion are transformed into each other and we assume the transformation matrix D to be irreducible. Now we introduce a perturbation V which has a symmetry lower than $\mathscr{H}$. This means that only a subgroup G_v of G will leave V invariant; the representation D_v of G_v must not be irreducible because D_v has a smaller dimension than D and may therefore be reducible. The number of times that each irreducible representation occurs in this reduction can be calculated with the help of formula (3.24).

Let us assume that D_v contains the two irreducible representations D_1

† This is equivalent to saying that there are degenerate energy levels in our system.

and D_2 with dimension n_1 and n_2 and $n_1 + n_2 = n$; we also assume that Ψ_{1i} $(i = 1, \ldots, n_1)$ are the n_1 eigenfunctions of D_1 and Ψ_{2j} $(j = 1, \ldots, n_2)$ are the n_2 eigenfunctions of D_2. Then we can calculate the three matrix elements

$$\begin{aligned} A_{ij} &= (\Psi_{1i}|V|\Psi_{2j}) \\ B_{ii'} &= (\Psi_{1i}|V|\Psi_{1i'}) \\ C_{jj'} &= (\Psi_{2j}|V|\Psi_{2j'}) \end{aligned} \tag{3.29}$$

By use of the orthogonality relations for the transformation matrix we find [AbBl70]

$$\begin{aligned} A_{ij} &= 0 \\ B_{ii'} &= \frac{\delta_{ii'}}{n_1} \sum_{k=1}^{n_1} (\Psi_{1k}|V|\Psi_{1k}) \\ C_{jj'} &= \frac{\delta_{jj'}}{n_2} \sum_{k=1}^{n_2} (\Psi_{2k}|V|\Psi_{2k}) \end{aligned} \tag{3.30}$$

These equations mean that the original $(n = 2J + 1)$-fold degenerate energy level is split into two energy levels, W_1 and W_2, of degeneracy n_1 and n_2 respectively; Ψ_{1i} and Ψ_{2j} are the correct zero order wavefunctions belonging to the irreducible representations D_1 and D_2. This reasoning can of course be applied if there are more than the two representations contained in D_v. These findings indicate how group theory is applied to crystal field theory. The only missing information is the character table for the group of three-dimensional rotations, which can be calculated from the equation [Ti64]

$$\chi_J(\alpha) = \frac{\sin (2J + 1)\alpha/2}{\sin \alpha/2} \tag{3.31}$$

for integral and nonintegral J, where α is the angle of rotation.

3.10 Example of Crystal Field Splitting: D_{3h} Symmetry

To demonstrate the foregoing general considerations we shall apply them to a case often encountered in the literature and in our examples, namely, D_{3h} symmetry (Fig. 9). The D_{3h} symmetry is formed from the D_3

symmetry by applying σ_h to each of its symmetry elements; this yields for D_{3h} the symmetry elements E, $2C_3$, $3C_2$, σ_h, $2S_3$, $3\sigma_v$. To determine the number of crystal field energy levels into which a term of total angular momentum J will be split, we call on the procedure outlined in the preceding sections. We therefore need the character table for D_{3h} symmetry for the different J values and symmetry elements that D_{3h} contains with respect to the rotation group R_3 [Be29].

Table 19 also gives the classification of the states with crystal quantum numbers, as obtained from Hellwege [He49]. The upper part of the character table is for terms with even J, whereas the lower part, containing the double groups, applies to terms with half integer J values. This character table shows an advantage of the group theoretical classification of the crystal states as compared to a crystal quantum number classification; the group theoretical classification is unique, whereas for some states two crystal quantum numbers are necessary.

TABLE 19

CHARACTER TABLE FOR D_{3h} SYMMETRY[a–f]

$\bar{\mu}$	$\bar{\nu}$	Γ	E	$2C_3$	$3C_2$	σ_h	$2S_3$	$3\sigma_v$	
0	0	$^1\Gamma_1$	1	1	1	1	1	1	
0	1	$^1\Gamma_2$	1	1	−1	1	1	−1	L_z
3	1	$^1\Gamma_3$	1	1	1	−1	−1	−1	
3	0	$^1\Gamma_4$	1	1	−1	−1	−1	1	z
±1		$^2\Gamma_5$	2	−1	0	−2	1	0	L_x, L_y
±2		$^2\Gamma_6$	2	−1	0	2	−1	0	x, y
±5/2		$2\Gamma_7$	2	1	0	0	$\sqrt{3}$	0	
±1/2		$^2\Gamma_8$	2	1	0	0	$-\sqrt{3}$	0	
±3/2		$^2\Gamma_9$	2	−2	0	0	0	0	

[a] The numbering of Γ_5 and Γ_6 follows Koster [KoDi63] and was adopted by Wybourne [Wy65]; Bethe [Be29] interchanged Γ_5 and Γ_6, a notation that has also been adopted by Hellwege [He49].

[b] In the original papers Hellwege [He49] used μ_t instead of $\bar{\mu}$ and S instead of $\bar{\nu}$ where $e^{i\bar{\nu}\pi} = S(\bar{\nu} = 0 \leftrightarrow S = 1; \bar{\nu} = 1 \leftrightarrow S = -1)$. In later papers Hellwege's group sometimes used ν instead of $\bar{\nu}$ [GrHe61].

[c] The classification of the $^2\Gamma_7$ and the $^2\Gamma_8$ state has been reversed with respect to Hellwege's labeling [He49].

[d] Hellwege [He49] gives $(\bar{\mu} = 3, \bar{\nu} = 0) = {}^1\Gamma_3$ and $(\bar{\mu} = 3, \bar{\nu} = 1) = {}^1\Gamma_4$.

[e] From [KoDi63, La74].

[f] L_x, L_y, L_z are the operators of the angular momentum; x, y, z those of the space coordinates.

TABLE 20

CHARACTERS OF THE D_{3h} SYMMETRY ELEMENTS WITH RESPECT TO THE R_3 GROUP[a]

	E	$2C_3$	$3C_2$	σ_h	$2S_3$	$3\sigma_v$
α:	0	$2\pi/3$	π	π	$\pi/3$	π
J			χ_J^b			
0	1	1	1	1	1	1
1	3	0	−1	−1	2	−1
2	5	−1	1	1	1	1
3	7	1	−1	−1	−1	−1
4	9	0	1	1	−2	1
5	11	−1	−1	−1	−1	−1
6	13	1	1	1	1	1
7	15	0	−1	−1	2	−1
8	17	−1	1	1	1	1

[a] From [Be29].
[b] $\chi_J = [\sin(2J+1)(\alpha/2)]/[\sin(\alpha/2)]$.

TABLE 21

CRYSTAL FIELD SPLITTING IN D_{3h} SYMMETRY

(a) Even electron systems		J:	0	1	2	3	4	5	6	7	8
$^1\Gamma_1$	$(\bar{\mu}=0, \bar{\nu}=0)$	p_{1J}	1	0	1	0	1	0	2	1	2
$^1\Gamma_2$	$(\bar{\mu}=0, \bar{\nu}=1)$	p_{2J}	0	1	0	1	0	1	1	2	1
$^1\Gamma_3$	$(\bar{\mu}=3, \bar{\nu}=1)$	p_{3J}	0	0	0	1	1	1	1	1	1
$^1\Gamma_4$	$(\bar{\mu}=3, \bar{\nu}=0)$	p_{4J}	0	0	0	1	1	1	1	1	1
$^2\Gamma_5$	$(\bar{\mu}=\pm 1)$	p_{5J}	0	1	1	1	1	2	2	3	3
$^2\Gamma_6$	$(\bar{\mu}=\pm 2)$	p_{6J}	0	0	1	1	2	2	2	2	3
Total number of levels			1	2	3	5	6	7	9	10	11

(b) Odd electron systems		J:	1/2	3/2	5/2	7/2	9/2	11/2	13/2	15/2
$^2\Gamma_7$	$(\bar{\mu}=\pm 5/2)$	p_{7J}	1	1	1	1	1	2	3	3
$^2\Gamma_8$	$(\bar{\mu}=\pm 1/2)$	p_{8J}	0	0	1	2	2	2	2	2
$^2\Gamma_9$	$(\bar{\mu}=\pm 3/2)$	p_{9J}	0	1	1	1	2	2	2	3
Total number of levels			1	2	3	4	5	6	7	8

TABLE 22

NUMBER OF CRYSTAL FIELD STATES FOR ALL SYMMETRIES[a]

Symmetry		J: 0 1 2 3 4 5 $\cdots$	p^b	J: $\frac{1}{2}$ $\frac{3}{2}$ $\frac{5}{2}$ $\frac{7}{2}$ $\frac{9}{2}$ $\frac{11}{2}$ $\cdots$	p^b
Triclinic Monoclinic Rhombic	$C_1, C_i,$ $C_s, C_2, C_{2h},$ C_{2v}, D_2, D_{2h}	1 3 5 7 9 11 $\cdots$ $2J + 1$	1		
Trigonal Hexagonal	$C_3, C_{3i}, C_{3v}, D_3, D_{3d},$ $C_{3h}, C_6, C_{6h}, D_{3h}, C_{6v}, D_6, D_{6h}$	1 1 1 3 3 3 $\cdots$ $2[J/3] + 1$[d] 0 1 2 2 3 4 $\cdots$ $[(2J - 1)/3] + 1$	1 2	1 2 3 4 5 6 $\cdots$ $J + (1/2)$ (Triclinic through Tetragonal)	2
Tetragonal	$S_4, C_4, C_{4h}, D_{2d}, C_{4v}, D_4, D_{4h}$	1 1 3 3 5 5 $\cdots$ $2[J/2] + 1$ 0 1 1 2 2 3 $\cdots$ $[(J^2 + 1)/2]$	1 2		
Cubic[c]	T, T_h, T_d, O, O_h	1 0 0 1 1 0 $\cdots$ $\varphi_1(x) + [J/6]$ 0 0 1 0 1 1 $\cdots$ $\varphi_2(x) + [J/6]$ 0 1 1 2 2 3 $\cdots$ $[(J + 1)/2]$	1 2 3	1 0 1 2 1 2 $\cdots$ $\varphi(y) + [J/3]$ 0 1 1 1 2 2 $\cdots$ $[(J + (3/2))/3]$	2 4

[a] From [FiJo57].

[b] p is the degeneracy of the level.

[c] The functions $\varphi_1(x)$, $\varphi_2(x)$ and $\varphi(y)$ are given by

x	0	1	2	3	4	5
$\varphi_1(x)$	1	0	0	1	1	0
$\varphi_2(x)$	0	0	1	0	1	1

y	$\frac{1}{2}$	$\frac{3}{2}$	$\frac{5}{2}$
$\varphi(y)$	1	0	1

and x and y are calculated for a given J by $J = x + 6\lambda$, $J = y + 3\lambda$, $(\lambda = 0, 1, 2, \ldots)$.

[d] The bracket indicates that in cases of a noninteger number the next smaller integer number should be taken and in cases of an integer number this should be taken.

With the information in Tables 19 and 20 we can calculate the number of times each representation occurs from

$$p_{sJ} = \frac{1}{n} \sum_{r=1}^{c} h_r \chi_J(\alpha_r) \chi_{sr}^* \tag{3.24a}$$

which yields results for even electron systems in Table 21. The analogous procedure can be performed for noninteger J values, yielding the results for odd electron systems in Table 21. In the manner outlined the number of crystal field states can be calculated for each J term. In Table 22 we summarize the results for all crystal symmetries [FiJo57].

TABLE 23

M_J COMPOSITION OF CRYSTAL FIELD STATES IN D_{3h} SYMMETRY

(a) Even number of electrons

Representation:	${}^1\Gamma_1$ $\bar{\mu} = 0$ $\bar{\nu} = 0$	${}^1\Gamma_2$ $\bar{\mu} = 0$ $\bar{\nu} = 1$	${}^2\Gamma_5$ $\bar{\mu} = \pm 1$	${}^2\Gamma_6$ $\bar{\mu} = \pm 2$	${}^1\Gamma_3$ $\bar{\mu} = 3$ $\bar{\nu} = 1$	${}^1\Gamma_4$ $\bar{\mu} = 3$ $\bar{\nu} = 0$
J	M_J	M_J	M_J	M_J	M_J	M_J
0	0	—	—	—	—	—
1	—	0	± 1	—	—	—
2	0	—	± 1	± 2	—	—
3	—	0	± 1	± 2	$+3, -3$	$+3, -3$
4	0	—	± 1	$\pm 2, \mp 4$	$+3, -3$	$+3, -3$
5	—	0	$\pm 1, \mp 5$	$\pm 2, \mp 4$	$+3, -3$	$+3, -3$
6	$+6, 0, -6$	$+6, -6$	$\pm 1, \mp 5$	$\pm 2, \mp 4$	$+3, -3$	$+3, -3$
7	$+6, -6$	$+6, 0, -6$	$\pm 1, \mp 5, \pm 7$	$\pm 2, \mp 4$	$+3, -3$	$+3, -3$
8	$+6, 0, -6$	$+6, -6$	$\pm 1, \mp 5, \pm 7$	$\pm 2, \mp 4, \pm 8$	$+3, -3$	$+3, -3$

(b) Odd number of electrons

Representation:	${}^2\Gamma_7$ $\bar{\mu} = \mp 5/2$	${}^2\Gamma_8$ $\bar{\mu} = \mp 1/2$	${}^2\Gamma_9$ $\bar{\mu} = \mp 3/2$
J	M_J	M_J	M_J
1/2	$\pm 1/2$	—	—
3/2	$\pm 1/2$	—	$\pm 3/2$
5/2	$\pm 1/2$	$\pm 5/2$	$\pm 3/2$
7/2	$\pm 1/2$	$\pm 5/2, \mp 7/2$	$\pm 3/2$
9/2	$\pm 1/2$	$\pm 5/2, \mp 7/2$	$\pm 3/2, \mp 9/2$
11/2	$\pm 1/2, \mp 11/2$	$\pm 5/2, \mp 7/2$	$\pm 3/2, \mp 9/2$
13/2	$\pm 1/2, \mp 11/2, \pm 13/2$	$\pm 5/2, \mp 7/2$	$\pm 3/2, \mp 9/2$
15/2	$\pm 1/2, \mp 11/2, \pm 13/2$	$\pm 5/2, \mp 7/2$	$\pm 3/2, \mp 9/2, \pm 15/2$

Finally, Table 23 shows explicitly the M_J composition of the various states (it is implicitly in Table 21), again for the case of D_{3h} symmetry. Also evident in Table 23 is a special feature of the crystal quantum number classification that is often overlooked. According to Hellwege [He49], the relation between the crystal quantum numbers $\bar{\mu}$ and M_J values is given by

$$\bar{\mu} = M_J \pm \frac{n}{2} \sum_i l_i \qquad (\text{mod } n) \tag{3.32}$$

where n is the multiplicity of the rotation axis for the symmetry under consideration (for $D_{3h}, n = 6$), l_i is the angular momentum of a 4f electron, and i is the number of 4f electrons for the particular RE ion; (mod n) means that we must choose the smallest number that can be reached by adding to or subtracting from $\bar{\mu}$ multiples of n. For even electron numbers, we find that for $J \leq 3$ the values of $\bar{\mu}$ and M_J correspond to each other (e.g., $M_J = 0 \leftrightarrow \bar{\mu} = 0$, etc.). For odd electron numbers and $J \leq \frac{5}{2}$, however, when $M_J = \pm\frac{1}{2}$ and $M_J = \pm\frac{5}{2}$, the crystal quantum number classification is reversed, so that $M_J = \pm\frac{1}{2}$ corresponds to $\bar{\mu} = \mp\frac{5}{2}$, etc. For larger J values Table 23 gives the appropriate classification accordingly.

3.11 Product of Two Representations: Selection Rules

If $\Gamma_i(A)$ and $\Gamma_j(A)$ are two representations of a group G, where A runs over all the elements of a group, it can be shown that the product

$$\Gamma(A) = \Gamma_i(A) \times \Gamma_j(A) \tag{3.33}$$

is also a representation of the group; this representation, however, is generally still reducible. We can express this by

$$\Gamma_i \times \Gamma_j = \sum_l a_{ijl} \Gamma_l \tag{3.34}$$

which means that if $\Gamma_i \times \Gamma_j$ is arranged in block form, the irreducible representation Γ_l occurs a_{ijl} times along the diagonal. The characters of $\Gamma_i \times \Gamma_j$ are given by

$$\chi = \chi_i \cdot \chi_j \tag{3.35}$$

These equations can be used to obtain selection rules by group theoretical techniques. (By this we mean rules that tell us whether the matrix elements of certain operators within a given basis set are zero or nonzero.) Let us consider a matrix element of the form

$$\langle \Psi_i | \mathscr{H}_j | \Phi_l \rangle = \int \Psi_i^* \mathscr{H}_j \Phi_l \, d\tau \tag{3.36}$$

where Ψ_i, $\mathscr{H}_j$, and Φ_l transform according to the irreducible representations Γ_i, Γ_j, and Γ_l of their group. (Assume real matrices, such that $\Gamma_i^* \equiv \Gamma_i$.) Then the matrix element can be different from zero only if $\Gamma_i \times \Gamma_j$ contains Γ_l at least once; since the equivalent statement holds for the characters χ_i, χ_j, and χ_l, the existence or nonexistence of a matrix element can be obtained from the character tables (which in turn can be found in many papers).

In an actual case, matters are simplified considerably because all the needed information is available in table form. The matrix elements of the electric dipole operators are those of (x, y, z) and of the magnetic dipole operators are, e.g., those of (L_x, L_y, L_z) (therefore their character is included in Table 19). Thus we need the representations of these operators in a given symmetry. They can be found, e.g., in Tinkham's book [Ti64, Appendix B]. Then the multiplication tables for the representation of a given group are needed, and they have been tabulated by Koster *et al.* [KoDi63]. To facilitate our understanding of the following example, we present the multiplication table for the representation of D_{3h} symmetry as Table 24. With this material, the selection rules are easily worked out, as will be shown in Section 3.12.

3.12 Example of Selection Rules: D_{3h} Symmetry

To find the selection rules for electric and magnetic dipole radiation for D_{3h} symmetry, we determine (e.g., from Tinkham [Ti64]) that z and x, y transform like ${}^1\Gamma_4$ and ${}^2\Gamma_6$ in D_{3h}, and L_z and L_x, L_y transform like ${}^1\Gamma_2$ and ${}^2\Gamma_5$, respectively. With Table 24 [KoDi63] we find

$$\begin{array}{lll} {}^1\Gamma_1 \times {}^1\Gamma_4 = {}^1\Gamma_4 & {}^1\Gamma_4 \times {}^1\Gamma_4 = {}^1\Gamma_1 & {}^2\Gamma_7 \times {}^1\Gamma_4 = {}^2\Gamma_8 \\ {}^1\Gamma_2 \times {}^1\Gamma_4 = {}^1\Gamma_3 & {}^2\Gamma_5 \times {}^1\Gamma_4 = {}^2\Gamma_6 & {}^2\Gamma_8 \times {}^1\Gamma_4 = {}^2\Gamma_7 \\ {}^1\Gamma_3 \times {}^1\Gamma_4 = {}^1\Gamma_2 & {}^2\Gamma_6 \times {}^1\Gamma_4 = {}^2\Gamma_5 & {}^2\Gamma_9 \times {}^1\Gamma_4 = {}^2\Gamma_9 \end{array} \tag{3.37}$$

From these equations the $E_{\|}$ polarization entries in Table 25, part a, are obtained. Here $E_{\|}$ denotes the occurrence of a transition in electric dipole radiation with polarization parallel to the crystal z axis. (The meaning of

TABLE 24

MULTIPLICATION TABLE FOR THE D_{3h} GROUP[a]

${}^1\Gamma_1$	${}^1\Gamma_2$	${}^1\Gamma_3$	${}^1\Gamma_4$	${}^2\Gamma_5$	${}^2\Gamma_6$	${}^2\Gamma_7$	${}^2\Gamma_8$	${}^2\Gamma_9$	
${}^1\Gamma_1$	${}^1\Gamma_2$	${}^1\Gamma_3$	${}^1\Gamma_4$	${}^2\Gamma_5$	${}^2\Gamma_6$	${}^2\Gamma_7$	${}^2\Gamma_8$	${}^2\Gamma_9$	${}^1\Gamma_1$
	${}^1\Gamma_1$	${}^1\Gamma_4$	${}^1\Gamma_3$	${}^2\Gamma_5$	${}^2\Gamma_6$	${}^2\Gamma_7$	${}^2\Gamma_8$	${}^2\Gamma_9$	${}^1\Gamma_2$
		${}^1\Gamma_1$	${}^1\Gamma_2$	${}^2\Gamma_6$	${}^2\Gamma_5$	${}^2\Gamma_8$	${}^2\Gamma_7$	${}^2\Gamma_9$	${}^1\Gamma_3$
			${}^1\Gamma_1$	${}^2\Gamma_6$	${}^2\Gamma_5$	${}^2\Gamma_8$	${}^2\Gamma_7$	${}^2\Gamma_9$	${}^1\Gamma_4$
				${}^1\Gamma_1 + {}^1\Gamma_2 + {}^2\Gamma_6$	${}^1\Gamma_3 + {}^1\Gamma_4 + {}^2\Gamma_5$	${}^2\Gamma_7 + {}^2\Gamma_9$	${}^2\Gamma_8 + {}^2\Gamma_9$	${}^2\Gamma_7 + {}^2\Gamma_8$	${}^2\Gamma_5$
					${}^1\Gamma_1 + {}^1\Gamma_2 + {}^2\Gamma_6$	${}^2\Gamma_8 + {}^2\Gamma_9$	${}^2\Gamma_7 + {}^2\Gamma_9$	${}^2\Gamma_7 + {}^2\Gamma_8$	${}^2\Gamma_6$
						${}^1\Gamma_1 + {}^1\Gamma_2 + {}^2\Gamma_5$	${}^1\Gamma_3 + {}^1\Gamma_4 + {}^2\Gamma_6$	${}^2\Gamma_5 + {}^2\Gamma_6$	${}^2\Gamma_7$
							${}^1\Gamma_1 + {}^1\Gamma_2 + {}^2\Gamma_5$	${}^2\Gamma_5 + {}^2\Gamma_6$	${}^2\Gamma_8$
								${}^1\Gamma_1 + {}^1\Gamma_2 + {}^1\Gamma_3 + {}^1\Gamma_4$	${}^2\Gamma_9$

[a] Reprinted from "Properties of the thirty-two point groups" by Koster *et al.* [KoDi63] by permission of the M.I.T. Press, Cambridge, Massachusetts.

TABLE 25

ELECTRIC DIPOLE SELECTION RULES FOR D_{3h} SYMMETRY[a]

(a) Even number of electrons

		${}^1\Gamma_1$ $\bar{\mu}=0$ $\bar{\nu}=0$	${}^1\Gamma_2$ $\bar{\mu}=0$ $\bar{\nu}=1$	${}^1\Gamma_3$ $\bar{\mu}=3$ $\bar{\nu}=1$	${}^1\Gamma_4$ $\bar{\mu}=3$ $\bar{\nu}=0$	${}^2\Gamma_5$ $\bar{\mu}=\pm 1$	${}^2\Gamma_6$ $\bar{\mu}=\pm 2$
$\bar{\mu}=0, \bar{\nu}=0$	${}^1\Gamma_1$	0	0	0	$E_\parallel$	0	$E_\perp$
$\bar{\mu}=0, \bar{\nu}=1$	${}^1\Gamma_2$	0	0	$E_\parallel$	0	0	$E_\perp$
$\bar{\mu}=3, \bar{\nu}=1$	${}^1\Gamma_3$	0	$E_\parallel$	0	0	$E_\perp$	0
$\bar{\mu}=3, \bar{\nu}=0$	${}^1\Gamma_4$	$E_\parallel$	0	0	0	$E_\perp$	0
$\bar{\mu}=\pm 1$	${}^2\Gamma_5$	0	0	$E_\perp$	$E_\perp$	$E_\perp$	$E_\parallel$
$\bar{\mu}=\pm 2$	${}^2\Gamma_6$	$E_\perp$	$E_\perp$	0	0	$E_\parallel$	$E_\perp$

(b) Odd number of electrons

		${}^2\Gamma_7$ $\bar{\mu}=\pm 5/2$	${}^2\Gamma_8$ $\bar{\mu}=\pm 1/2$	${}^2\Gamma_9$ $\bar{\mu}=\pm 3/2$
$\bar{\mu}=\pm 5/2$	${}^2\Gamma_7$	0	$E_\perp, E_\parallel$	$E_\perp$
$\bar{\mu}=\pm 1/2$	${}^2\Gamma_8$	$E_\perp, E_\parallel$	0	$E_\perp$
$\bar{\mu}=\pm 3/2$	${}^2\Gamma_9$	$E_\perp$	$E_\perp$	$E_\parallel$

[a] $E_\parallel$ ($E_\perp$) means that the incident radiation is polarized with its electric vector parallel (perpendicular) to the crystal axis.

TABLE 26

MAGNETIC DIPOLE SELECTION RULES FOR D_{3h} SYMMETRY[a]

(a) Even number of electrons (and $\Delta J = 0, \pm 1$, $J = 0 \to J = 0$ forbidden)

		$^1\Gamma_1$ $\bar{\mu} = 0$ $\bar{\nu} = 0$	$^1\Gamma_2$ $\bar{\mu} = 0$ $\bar{\nu} = 1$	$^1\Gamma_3$ $\bar{\mu} = 3$ $\bar{\nu} = 1$	$^1\Gamma_4$ $\bar{\mu} = 3$ $\bar{\nu} = 0$	$^2\Gamma_5$ $\bar{\mu} = \pm 1$	$^2\Gamma_6$ $\bar{\mu} = \pm 2$
$\bar{\mu} = 0, \bar{\nu} = 0$	$^1\Gamma_1$	0	$M_\parallel$	0	0	$M_\perp$	0
$\bar{\mu} = 0, \bar{\nu} = 1$	$^1\Gamma_2$	$M_\parallel$	0	0	0	$M_\perp$	0
$\bar{\mu} = 3, \bar{\nu} = 1$	$^1\Gamma_3$	0	0	0	$M_\parallel$	0	$M_\perp$
$\bar{\mu} = 3, \bar{\nu} = 0$	$^1\Gamma_4$	0	0	$M_\parallel$	0	0	$M_\perp$
$\bar{\mu} = \pm 1$	$^2\Gamma_5$	$M_\perp$	$M_\perp$	0	0	$M_\parallel$	$M_\perp$
$\bar{\mu} = \pm 2$	$^2\Gamma_6$	0	0	$M_\perp$	$M_\perp$	$M_\perp$	$M_\parallel$

(b) Odd number of electrons (and $\Delta J = 0, \pm 1$)

		$^2\Gamma_7$ $\bar{\mu} = \pm 5/2$	$^2\Gamma_8$ $\bar{\mu} = \pm 1/2$	$^2\Gamma_9$ $\bar{\mu} = \pm 3/2$
$\bar{\mu} = \pm 5/2$	$^2\Gamma_7$	$M_\parallel, M_\perp$	0	$M_\perp$
$\bar{\mu} = \pm 1/2$	$^2\Gamma_8$	0	$M_\parallel, M_\perp$	$M_\perp$
$\bar{\mu} = \pm 3/2$	$^2\Gamma_9$	$M_\perp$	$M_\perp$	$M_\parallel$

[a] Note that the electric and magnetic vectors of the radiation are perpendicular to each other; since the polarization of light is usually defined with respect to the electric vector, $M_\parallel$ corresponds to σ-polarized light and $M_\perp$ to π-polarized light.

$E_\perp$ is analogous.) The rest of the selection rules in Tables 25 and 26 are produced in a similar manner. To facilitate reading the literature, the crystal quantum number classification of the levels is included. In this notation the following selection rules hold [He49]

$$\text{Electric dipole radiation:} \quad \begin{array}{lll} E_\perp \neq 0 & \text{for} & \Delta\bar{\mu} = \pm 2, \pm 4, \ldots. \\ E_\parallel \neq 0 & \text{for} & \Delta\bar{\mu} = \pm 3 \\ & \text{and} & \Delta\bar{\nu} = 0 \end{array} \tag{3.38}$$

($\Delta\bar{\nu} = 0$ necessary only for even number electron system)

$$\text{Magnetic dipole radiation}\dagger\text{:} \quad \begin{array}{lll} M_\perp \neq 0 & \text{for} & \Delta\bar{\mu} = \pm 1, \pm 5, \ldots \\ M_\parallel \neq 0 & \text{for} & \Delta\bar{\mu} = 0 \\ & \text{and} & \Delta\bar{\nu} = \pm 1 \end{array} \tag{3.39}$$

† In the magnetic dipole radiation case we have to realize that if a matrix element for L_z is nonzero, this corresponds to σ-polarized light, because the polarization of the radiation is generally given with respect to its electric vector!

($\Delta\bar{\nu} = \pm 1$ necessary only for even number electron system) and $\Delta J = 0$, ± 1 ($J = 0 \rightarrow J = 0$ forbidden).

The latter selection rules are also important for analyzing the Zeeman effect patterns because the same operator is effective there. The selection rules show, e.g., that, as expected, all the doublet levels show a linear Zeeman effect for a magnetic field parallel to the principal axis; a linear Zeeman effect for a magnetic field perpendicular to the principal axis may be found only for the $^2\Gamma_8$ and the $^2\Gamma_7$ states, because only they can have nonzero matrix elements with $\Delta M_J = \pm 1$.

D_{3h} symmetry is a useful subject for a brief comparison of the group theoretical classification [Be29] and the crystal quantum number classification [He49] in crystal spectra. Group theory is a very general tool in physics and is therefore widely known and can be easily applied to analyze crystal spectra because almost all the necessary information is now available in table form. Differences in the notation used by different authors constitute a problem. Moreover, Kramers degeneracy introduces some complications into the group theoretical approach, although again the difficulty is hardly noticed as long as we can rely on tabular material. The crystal quantum number concept is very appealing because it follows the lines of atomic spectroscopy, with which we are familiar from basic quantum mechanics. Also, in the crystal quantum number approach the selection rules can be written in a concise form. A problem, however, is that in some (although only a few) cases (D_{3h} is one of them) two crystal quantum numbers are necessary, which may complicate matters somewhat. Also, in the spectroscopy of 3d ions in crystals this approach has hardly been used at all. Therefore, the group theoretical concept is more widely applicable, although for a quick spectral analysis the crystal quantum number approach is handier. We shall present both types of classification in parallel, in order to facilitate the reading of the original literature.

3.13 Application of Theoretical Results to the Analysis of Experimental Data

The foregoing considerations necessarily seem quite abstract and their use is certainly not apparent. Therefore we give a few examples of their application, starting with the spectrum used to outline the problems in this field at the beginning of the chapter (Fig. 8). It shows the $^4I_{15/2} \rightarrow {}^4S_{3/2}$ transition of Er in $La(C_2H_5SO_4)_3 \cdot 9H_2O$, a crystal of D_{3h} symmetry

[FiRu59, Ke37]. Assume for the moment that we do not know the assignment of the excited term. We see from Table 22 that the J value of the excited term can be determined from the number of lines in crystal spectra whose symmetry is less than cubic, because it is always $(2J + 1)/2$ for a system with an odd number of 4f electrons.

We must keep in mind, however, that due to the selection rules the number of transitions may be smaller than the number of crystal field levels in the excited term. For an unambiguous determination of the J value it is therefore always advisable to use crystals with low symmetry, in which there are only few selection rules.

In the present case we find that the transition always consists of two lines,† meaning that the excited term has $J = \frac{3}{2}$. According to Table 21, part b, this term is split by the crystal field into two states with the classifications $^2\Gamma_7$ ($\bar{\mu} = \pm\frac{5}{2}$) and $^2\Gamma_9$ ($\bar{\mu} = \pm\frac{3}{2}$). The lowest crystal field level of the ground term can now have the classification $^2\Gamma_7$, $^2\Gamma_8$, or $^2\Gamma_9$. Using the selection rules (Table 25), we find the following possible transitions for the three possible assignments‡):

$^2\Gamma_7$: σ component (to $^2\Gamma_9$)

$^2\Gamma_8$: σ component (to $^2\Gamma_9$), π/σ component (to $^2\Gamma_7$)

$^2\Gamma_9$: π component (to $^2\Gamma_9$), σ component (to $^2\Gamma_7$)

The transition in Fig. 8 consists of one σ component and one π/σ component, giving the classification $^2\Gamma_8$ for the ground state and the classifications $^2\Gamma_9$ for the higher energy crystal field level of the excited term and $^2\Gamma_7$ for the lower crystal field level. The ground state classification can now be used to analyze spectra obtained from transitions to other excited terms in $Er(C_2H_5SO_4)_3 \cdot 9H_2O$, and in no case does the classification just obtained lead to any inconsistencies. We might add that the given ground state classification is consistent with that obtained by ESR investigation [AbBl70].

As another example we use the $^7F_0 \rightarrow {}^5D_0$, $^5D_1$, and $^5D_2$ transitions in $Eu(C_2H_5SO_4)_3 \cdot 9H_2O$ [HeJo57]. The selection rules in Table 25 tell us that the $^7F_0 \rightarrow {}^5D_0$ transition is forbidden, the $^7F_0 \rightarrow {}^5D_1$ transition shows only magnetic dipole radiation, and the $^7F_0 \rightarrow {}^5D_2$ transition shows only

† We tacitly assume that the experiments are carried out at low temperatures (4.2 K) such that only the lowest crystal field level of the ground term is occupied and the number of lines directly reflects the number of crystal field levels in the excited term.

‡ Since the ground term is $^4I_{15/2}$ and the excited term has $J = \frac{3}{2}$, the observed transitions must be electric dipole in nature.

electric dipole radiation. More specifically, the $^7F_0$ ground state is $^1\Gamma_1$ and has these transitions to the $^5D_1$ term in magnetic dipole radiation:

$$^7F_0 \rightarrow {}^5D_1\!: \quad {}^1\Gamma_1 \rightarrow {}^1\Gamma_2\!: \quad \sigma \quad (M_z \neq 0)$$

$$^1\Gamma_1 \rightarrow {}^2\Gamma_5\!: \quad \pi \quad (M_x \pm iM_y \neq 0)$$

(Here the polarization is given with respect to the electric vector of the radiation, which is perpendicular to the magnetic vector; the nonvanishing components of the magnetic dipole operator are given in parentheses.) Indeed, the spectrum shows two polarized lines, the π-polarized line having the lower energy (Fig. 12a). The correctness of the classification can also be seen from the Zeeman effect measurements (for details see Section 3.15). The singlet $^1\Gamma_2$ is not influenced by a magnetic field, whereas the $^2\Gamma_5$ doublet is split by a magnetic field, as can be seen in Fig. 12a.

This example can also be used to demonstrate how to distinguish experimentally between electric and magnetic dipole radiation. For this purpose we show in Fig. 12b the $^7F_0 \rightarrow {}^5D_1$ transition; in contrast to Fig. 12a, however, the orientation of the crystal with respect to the direction of the magnetic field has been changed. Let us assume that we do not know whether the transition $^7F_0 \rightarrow {}^5D_1$ consists of magnetic or electric dipole radiation. In addition, we limit ourselves to the spectra with zero external field. Light is directed at the crystal with polarization (measured with respect to the electrical vector) parallel and perpendicular to the direction into which the magnetic field can be applied according to the figure. In the configuration at the left in Fig. 12b we see one line with π polarization and one with σ polarization. The crystal is then rotated 90°, so that the propagation direction of the light (**S**) is now parallel to the crystal z axis. In the case of magnetic dipole radiation, only the π-polarized line observed before will occur because we can now excite only a magnetic vector perpendicular to the crystal axis (it does not matter how the light is polarized). This case corresponds to the actual situation. If the transition were an electric dipole transition, the σ polarized line observed before would now appear. The identification of the lines can be checked by comparing the spectra with the magnetic field, which in each case acts perpendicular to the crystal axis. This shows that the $H = 0$ line in the spectrum at the right is the π-polarized line in the spectrum at the left.

The magnetic field destroys the symmetry of the system and therefore upon its application the selection rules are relaxed making all lines in the finite external field spectra visible.

Finally, the selection rules tell us that the $^7F_0 \rightarrow {}^5D_2$ transition at the right in Fig. 12a should consist of one σ-polarized transition in electric

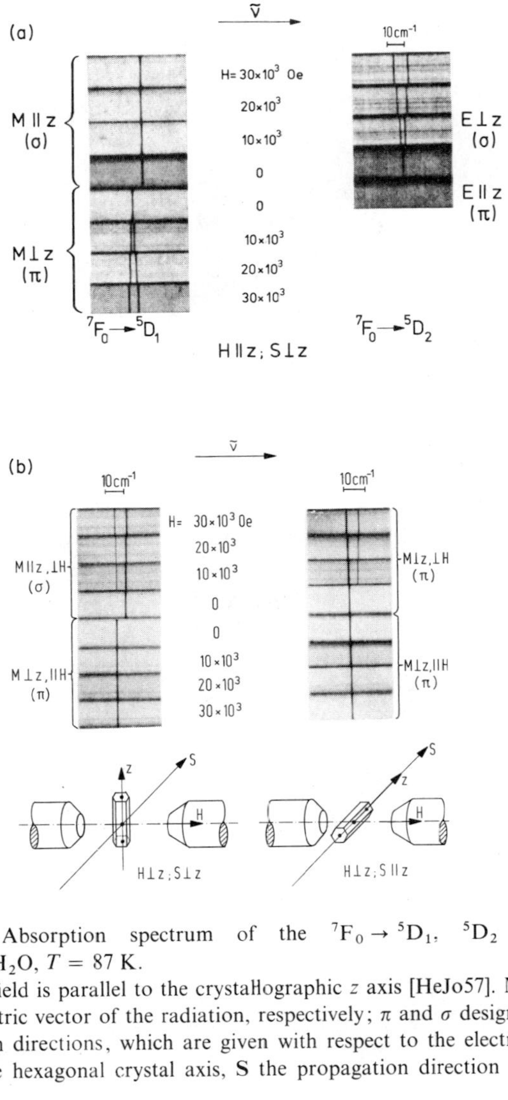

Fig. 12. (a) Absorption spectrum of the $^7F_0 \rightarrow {}^5D_1$, $^5D_2$ transitions in $Eu(C_2H_5SO_4)_3 \cdot 9H_2O$, $T = 87$ K.

The magnetic field is parallel to the crystallographic z axis [HeJo57]. **M** and **E** are the magnetic and electric vector of the radiation, respectively; π and σ designate the conventional polarization directions, which are given with respect to the electric vector of the radiation. z is the hexagonal crystal axis, **S** the propagation direction of the radiation (pointing vector).

(b) Absorption spectrum of the $^7F_0 \rightarrow {}^5D_1$ transition in $Eu(C_2H_5SO_4)_3 \cdot 9H_2O$, $T = 87$ K.

The symbols have the same meaning as in Fig. 12a. The orientation of the crystal has been changed with respect to that in Fig. 12a, so as to permit us to determine the nature of the radiation (magnetic dipole) and to show the Zeeman effect perpendicular to the crystal axis [HeJo57].

dipole radiation ($^1\Gamma_1 \to {}^2\Gamma_6$). This is indeed observed and the Zeeman effect proves that it is a twofold degenerate level with twice the splitting factor shown by the ($^1\Gamma_1 \to {}^2\Gamma_5$) transition for $^5F_0 \to {}^5D_1$.

3.14 The Zeeman Effect

Our discussion of crystal field splitting has shown that for RE ions with an odd number of electrons every crystal field level is twofold degenerate (Kramers degeneracy) and that for ions with an even number of electrons twofold levels are possible, depending on the symmetry. In this chapter we ignore fourfold degenerate levels because they can occur only in cubic symmetry and have seldom been observed in optical spectra. In ESR investigations, however, their role is significant, and they are dealt with in detail by Abragam and Bleaney [AbBl70]. The degeneracy of these twofold levels can be removed by an externally applied magnetic field. The splitting observed in the spectra makes possible the determination of the magnetic moments of the crystal field levels involved in the transition. The Hamiltonian for the Zeeman interaction has the general form

$$\mathcal{H}^{Z} = \mu_B \mathbf{H} \cdot (\mathbf{L} + 2\mathbf{S}) \tag{3.40}$$

where **H** is the vector of the magnetic field.

Except for a few cases, it is usually a good approximation to assume that the Zeeman interaction is small compared to the crystal field interaction. (There are, however, notable exceptions, an example of which will be given later.) To analyze the Zeeman interaction we start with the wavefunctions of the crystal field levels. If the Hamiltonian (3.40) is used to calculate the Zeeman effect of a crystal field level we find that the Zeeman energy is anisotropic—which is a consequence of the anisotropy introduced by the crystal field interaction. This anisotropic Zeeman splitting can be described by an anisotropic g-factor (g-tensor). Its principal axis do not necessarily coincide with those of the crystal coordinate system. In dealing with the Zeeman interaction it is, however, often convenient to work in a coordinate system that coincides with the directions of the principal axes of the g-tensor. From measurements in the three principal axes the Zeeman effect in an arbitrary direction can be easily predicted. In an uniaxial crystal we distinguish the Zeeman effect parallel to the crystal axis and call it the parallel Zeeman effect (in this direction the wavefunctions are diagonal in M_J) and the Zeeman effect perpendicular to it which we call the perpendicular Zeeman effect. Thus two kinds of matrix elements must be distinguished

$$\text{Parallel Zeeman effect:} \quad (J\Gamma|(L_z + 2S_z)|J\Gamma')$$

and (3.41)

$$\text{Perpendicular Zeeman effect:} \quad (J\Gamma|(L_x + 2S_x)|J\Gamma')$$

Here we have assumed that J is a good quantum number for a particular crystal field level (no J mixing). Γ and Γ' are the representations of the two levels connected by the Zeeman operator. To determine for which states these matrix elements are nonzero, it is most convenient to use the character tables. In principle, the same selection rules are active here as for the magnetic dipole radiation, because we are dealing with the same operator. For practical purposes the following simple considerations are helpful. For a crystal field state containing only one $\pm M_J$ we have

$$(JM_J|\mathscr{H}_{\parallel}^{z}|JM_J') = \mu_B g_J H_{\parallel} M_J \delta_{M_J M_J'} \tag{3.42}$$

where g_J is the Landé g_J-factor. If a state consists of a linear combination of M_J values

$$|\;) = a_1|JM_{J1}) + a_2|JM_{J2}) + \cdots$$

we obtain

$$(J\Gamma|\mathscr{H}_{\parallel}^{z}|J\Gamma') = \mu_B g_J H_{\parallel} [|a_1^2|M_{J1} + |a_2^2|M_{J2} + \cdots] \tag{3.43}$$

It is convenient to define the splitting factor $s_{\parallel}$ as

$$s_{\parallel}(J\Gamma) = \mu_B g_J [|a_1^2|M_{J1} + |a_2^2|M_{J2} + \cdots] \tag{3.44}$$

In paramagnetic resonance notation [AbBl70] this quantity is half of the parallel g-factor $g_{\parallel}$ (for effective spin $S = \frac{1}{2}$). Thus for doubly degenerate crystal levels there is always a linear Zeeman effect, where the magnitude of the splitting is given by M_J values weighted by the squares of the coefficients of their wavefunctions. For the perpendicular Zeeman effect we have to realize that it can only occur if the Zeeman operator acts between states having M_J values that differ by ± 1, because

$$(JM_J|\mathscr{H}_{\perp}^{z}|JM_J \pm 1) = \tfrac{1}{2} g_J \mu_B H_{\perp}[(J \mp M_J)(J \pm M_J + 1)]^{1/2} \tag{3.45}$$

Thus many twofold degenerate levels will not show a linear Zeeman effect in a perpendicular field. As in the case of the parallel Zeeman effect, we can define the splitting factor

$$s_{\perp} = \tfrac{1}{2} g_J \mu_B [(J \mp M_J)(J \pm M_J + 1)]^{1/2} \tag{3.46}$$

which in paramagnetic resonance notation is written as $\frac{1}{2}g_\perp$ (for effective spin $\bar{S} = \frac{1}{2}$).

To illustrate this section we choose the Zeeman effect of a $J = 1$ state in D_{3h} symmetry; this arises, e.g., for the $^5D_1$ term in $Eu(C_2H_5SO_4)_3 \cdot 9H_2O$, as well as for the system of the two lowest crystal field levels in the ground term $^5I_8$ of Ho in $Ho(C_2H_5SO_4)_3 \cdot 9H_2O$, which has been treated extensively by Baker and Bleaney [BaBl58]. In the latter case we are not really dealing with a $J = 1$ state, but rather with a doublet and a singlet, near each other but remote from the other crystal field levels of the $^5I_8$ ground term, which can therefore be described by an effective spin of $\bar{S} = 1$.

For a state with $J = 1$ and a hexagonal crystal field, the secular equation can be written as (this is also valid for tetragonal or trigonal symmetry)

$$\begin{vmatrix} \frac{1}{3}D + g_J\mu_B H_\parallel - W & -\frac{1}{\sqrt{2}} g_J \mu_B H_\perp & 0 \\ -\frac{1}{\sqrt{2}} g_J\mu_B H_\perp & -\frac{2}{3}D - W & -\frac{1}{\sqrt{2}} g_J\mu_B H_\perp \\ 0 & -\frac{1}{\sqrt{2}} g_J\mu_B H_\perp & \frac{1}{3}D - g_J\mu_B H_\parallel - W \end{vmatrix} = 0$$

where $H_\parallel$ and $H_\perp$ are the magnetic fields parallel and perpendicular to the crystal axis and $-D$ is the crystal field splitting between the $M_J = 0$ and the $M_J = \pm 1$ state. From this equation we obtain for the magnetic field parallel to the hexagonal axis

$$\begin{aligned} W_0 &= -\tfrac{2}{3}D \\ W_\pm &= \tfrac{1}{3}D \pm g_J\mu_B H_\parallel \end{aligned} \tag{3.47a}$$

yielding, as expected, a field-independent behavior for the $M_J = 0$ state and a linear Zeeman splitting for the $M_J = \pm 1$ states. For the magnetic field perpendicular to the crystal axis, $H_\perp > 0$, we obtain the energy levels

$$\begin{aligned} W_{+1} &= \tfrac{1}{3}D \\ W_{\substack{0\\-1}} &= -\frac{1}{6}D \mp \frac{D}{2}\left(1 + \left(\frac{4g_J\mu_B H_\perp}{2D}\right)^2\right)^{1/2} \end{aligned} \tag{3.47b}$$

Note that for the $^5D_1$ term we have $D = -4.6\ \mathrm{cm}^{-1}$. The indices $+1, 0, -1$ correspond to the actual M_J values only for zero field. At finite magnetic fields the wavefunctions of the states are mixtures of the zero field wavefunctions. This produces a nonlinearity in the Zeeman pattern for weak fields, whereas at strong fields a linear Zeeman effect is observed again.

These Zeeman patterns can be seen in Figs. 12a and 12b. The left part of Fig. 12a shows the parallel Zeeman effect, where the ${}^2\Gamma_5$ ($\bar{\mu} = \pm 1$) level reveals a linear Zeeman effect (note that we are actually measuring transition energies; since the ${}^7F_0$ (${}^1\Gamma_1$) ground state is not affected by the magnetic field, the transition energies show the behavior of the ${}^5D_1$ term in the magnetic field). Similarly, Fig. 12b gives the perpendicular Zeeman effect. Figure 13 shows the energies as a function of the magnetic field, where the points have been measured and the solid curves are from eqs. (3.47a) and (3.47b). The broken lines for the perpendicular Zeeman effect are the strong field asymptotes.

From the Zeeman patterns for $H_{\parallel}$ and $H_{\perp}$, g_J can be determined. In both cases we obtain $g_J = 1.49$, in comparison to the Russell–Saunders value of 1.50, which shows that the intermediate coupling mixing is weak.

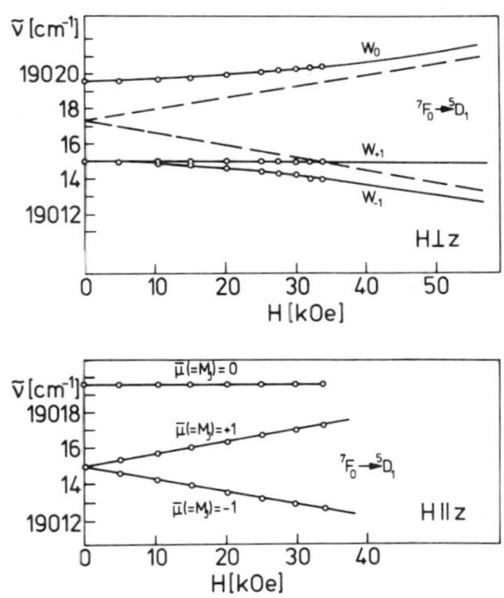

Fig. 13. Zeeman effect of the ${}^7F_0 \rightarrow {}^5D_1$ transition in $Eu(C_2H_5SO_4)_3 \cdot 9H_2O$ at $T = 87$ K, taken from the spectra in Figs. 12a and 12b.

(a) (Upper part) magnetic field H perpendicular to the hexagonal axis. Data are from Fig. 12b, where W_{+1} comes from the spectra $\mathbf{M} \perp z$ ($\mathbf{S} \perp z$) or $\mathbf{M} \perp z$, ($\mathbf{S} \parallel z$) and W_0 and W_{-1} come from the spectra $\mathbf{M} \parallel z$ ($\mathbf{S} \perp z$) or $\mathbf{M} \perp z$, ($\mathbf{S} \parallel z$).

(b) (Lower part) magnetic field H parallel to the hexagonal axis. The classification of the states is with crystal quantum numbers $\bar{\mu}$ (which for this case equal the M_J values). Data are from Fig. 12a [HeJo57].

Extensive experimental and theoretical Zeeman effect studies on RE ethylsulfates have been performed by Murao *et al.* [MuSp65, MuHa67, SyHa68, KaHa72a]. In this work also the angular variation of Zeeman patterns was investigated.

For even electron ions Griffith [Gr63b] has proven a very interesting and useful theorem. If in such a system there is a doublet allowed by symmetry that is so remote from other crystal field levels that second order effects can be neglected, its g-tensor, if brought to the principal axes, always has only one nonzero component, where the direction of this component is given by the crystal symmetry. Thus if in such a situation the one principal value of the g-tensor, which we can call g_z, has been measured, g_x and g_y have to be zero. These statements are no contradiction to the $J = 1$ case just treated, where also for $H_\perp$ a nonzero Zeeman effect was observed, because here the Zeeman splitting for $H_\perp$ is produced by a second order effect between the $|0\rangle$ and $|\pm 1\rangle$ states; thus the $|\pm 1\rangle$ state is not far removed from the $|0\rangle$ state, which is required for Griffith's theorem.

3.15 Intensities

The question of what intensity a particular transition between two crystal field levels has is one that can be broken up into two aspects. First, we have to work out the selection rules, which are imposed by the symmetry of the crystal; they can be obtained by the application of group theory, as shown in Section 3.12. Second, we have to ascertain which selection rules of the free RE ion "survive" the transformation of the RE ion from the gas into the crystal, and how they may be modified by the properties of the crystal. This quite important second aspect was dealt with systematically for the first time by van Vleck [vVl 37]. All the important points set down in this work are still valid, and have since been brought into a more mathematical form [Ju62, Of62].

The experimental data on the solid state spectra of RE show that the radiation is mostly electric dipole in nature, though some cases of magnetic dipole radiation are also observed. Since the optical transitions take place between levels of a particular $4f^N$ configuration, the electric dipole radiation is forbidden in first order, because the electric dipole operator has uneven parity and the transition matrix element must have even parity (Laportes selection rule). Van Vleck [vVl 37] pointed out that electric dipole radiation can only occur because the $4f^N$ states have admixtures of

$4f^{N-1}nl$ configurations (nl will be mostly 5d), where $4f^{N-1}nl$ has to be chosen such that it has opposite parity from $4f^N$. The admixture of the $4f^{N-1}nl$ wavefunctions into the $4f^N$ wavefunctions is produced by interactions that have odd parity. In the crystal there are two mechanisms that can produce these admixtures, namely odd parity crystal field components and crystal vibrations of odd symmetry. We note that in crystals where there is an inversion center, the odd crystal field components are zero, meaning that there can be no electric dipole radiation induced by the crystal field. Of course, the possibility of inducing it by odd symmetry vibrations remains. The four dominant sources of optical radiation in RE crystal spectra are

(I) forced electric dipole radiation induced by odd terms of the crystal field;
(II) forced electric dipole radiation induced by lattice vibrations;
(III) allowed magnetic dipole radiation;
(IV) allowed electric quadrupole radiation.

In free atoms, magnetic dipole radiation is about six orders of magnitude weaker than electric dipole radiation; thus since the latter occurs only as a consequence of a perturbation, both kinds of radiation show up in the RE spectra with about the same intensity; so far, quadrupole radiation has not been observed in RE crystal spectra. With the advent of tunable lasers (dye lasers), however, it may be possible to induce quadrupole transitions.

Intra-4f transitions can be brought about also by using the electronic Raman effect. The Raman operator is an even parity operator (consisting of two electric dipole operators); thus transitions between states of the same parity are allowed, though the transition probabilities are small because the Raman effect is a second order effect. A number of such investigations have, however, been performed (see, e.g., [KoScha70, KoMo73]).

The two-photon transitions belong in the same class as those induced by the Raman effect. The two-photon operator has even parity and therefore has allowed intra-4f matrix elements. The transition probability is again weak, and two-photon absorption experiments are only possible with lasers. So far, a two-photon intra-4f transition has been observed only for $Eu^{2+}:CaF_2$ and $Eu^{2+}:SrF_2$ [FrScha76]. (Two-photon absorption in $Eu^{2+}:CaF_2$ by the $4f^7(^8S_{7/2}) \rightarrow 4f^65d$ transition had been observed before [KaGa61b, BaScha70]; in these experiments the transition is forbidden for the two-photon operator, hence an odd symmetry phonon also has to be excited in order to preserve parity.)

For the first order allowed magnetic dipole radiation the free atom selection rules are still quite valid: $\Delta J = 0, \pm 1$ ($J = 0 \leftrightarrow J = 0$ forbidden) and $\Delta S = 0, \Delta L = 0$, where the latter two selection rules are broken because the wavefunctions are not Russell–Saunders but intermediate coupling wavefunctions. In the electric dipole case, however, because it is induced by the crystal field, the free atom selection rules break down almost completely, and the selection rules with regard to the L, S, J quantum numbers are now essentially governed by the crystal field interaction yielding $\Delta J \leq 6$ (approximately).

We now outline the Judd–Ofelt theory for the parametrization of the intensities in RE crystal spectra. It is in essence a quantification of the ideas formulated by van Vleck [vVl 37] under optical radiation sources (I) and (II). We have to calculate the matrix elements of the electric dipole moment, which is written in tensor operator form as [Wy65, Pe75, Re73, Re75]

$$E_p = -e \sum_i (r_i)_p = -er \sum_i (C_{1p})_i \tag{3.48}$$

where $p = 0$ gives the z component of the dipole and $p = \pm 1$ gives the $x \pm iy$ components of the dipole. As mentioned, we shall assume that the other parity φ_{nl} states are mixed into the 4f states $|4f^N JM_J)$ by the crystal field in order to produce nonvanishing matrix elements of the electric dipole operator, which means we have states

$$|a) = |f^N JM_J) + \sum_{\varphi_{nl}} \frac{(f^N JM_J|V|\varphi_{nl})|\varphi_{nl})}{W(f^N JM_J) - W(\varphi_{nl})}$$

where V is the operator of the crystal field interaction energy (see eq. (3.2)). It should be pointed out that now only the crystal field terms with odd k give nonzero matrix elements, because the two wavefunctions in the matrix elements have different parity.

The matrix element of the electric dipole operator between states a and a' is given by (neglecting higher order contributions)

$$\begin{aligned}(a|E_p|a') = &\sum_{\varphi_{nl}} \frac{(f^N JM_J|V|\varphi_{nl})(f^N J'M'_J|E_p|\varphi_{nl})}{W(f^N JM) - W(\varphi_{nl})}\\ &+ \sum_{\varphi_{nl}} \frac{(f^N J'M'_J|V|\varphi_{nl})(f^N JM_J|E_p|\varphi_{nl})}{W(f^N J'M'_J) - W(\varphi_{nl})}\end{aligned} \tag{3.49}$$

In order to evaluate this equation further, two drastic assumptions have to be made: (I) It is assumed that the states φ_{nl} are degenerate with respect to all their quantum numbers (except n and l). If we assume that φ_{nl} is φ_{5d}, we know from Fig. 6 that the energies of this configuration extend from about 50,000 to at least 100,000 cm^{-1}. This range indicates that this assumption imposes considerable strain on the final result; it has the virtue, however, that the summation over φ_{nl} can be carried out ("closure"), making the final result much more tractable. (II) It is assumed that

$$W(f^N J M_J) - W(\varphi_{nl}) = W(f^N J' M'_J) - W(\varphi_{nl}) \tag{3.50}$$

This assumption is a drastic one because the errors introduced by taking, e.g., a transition from a term at 25,000 cm^{-1} to the ground term, with the center of the 5d configuration assumed to be at 70,000 cm^{-1}, are not negligible. This approximation, however, makes it possible to combine the two sums in eq. (3.49) into one, in which only three terms are nonzero.

Evaluation of eq. (3.49) by use of the foregoing assumption involves the application of tensor operator algebra. The result is

$$\begin{aligned}(a|E_p|a') = -e \sum_{k,\,q,\,\lambda=\text{even}} & (-1)^{p+q+J-M_J} B^*_{kq}(2\lambda+1) \\ & \times Y_{k\lambda}\begin{pmatrix} 1 & \lambda & k \\ p & -q-p & q \end{pmatrix}\begin{pmatrix} J & \lambda & J' \\ -M_J & +q+p & M'_J \end{pmatrix} \\ & \times (f^N J \| U_\lambda \| f^N J')\end{aligned} \tag{3.51}$$

(where the asterisk means that for B^*_{kq} the radial integral is to be taken between states 4f and nl instead of 4f and 4f) with

$$\begin{aligned}Y_{k\lambda} = 2 \sum_{n,l} & \begin{Bmatrix} 1 & \lambda & k \\ f & l & f \end{Bmatrix} (f\|C_1\|l)(f|r|nl) \\ & \times (f\|C_k\|l)\cdot(W(f^N J M_J) - W(\varphi_{nl}))^{-1}\end{aligned} \tag{3.52}$$

(where f stands for the angular momentum of the f electrons, namely 3).

In this form eq. (3.51) can be used to calculate the intensity between individual crystal field levels, although doing so is a considerable task. It is better to deal with simpler cases to test the theory. We can, for example, compare the total absorption or emission intensities between terms, thus getting rid of all the summations that depend on the directions in space. The resulting equations are also applicable for the study of solution spectra, where the first systematic measurements of intensities were in fact made [BrGo45].

Intensities are often expressed in terms of oscillator strengths, where the oscillator strength for a transition from crystal field level a to crystal field level a' is given as [Ju63]

$$f_{aa'}^{\mathrm{ed}} = \frac{8\pi^2 m\nu}{he^2}\chi(a|E_p|a')^2 \tag{3.53}$$

where m is the electron mass, ν the frequency of the transition $a \rightarrow a'$, $\chi = n(n^2+2)^2/9$, a correction for the refractive index n of the crystal. In order to be applicable to transitions between terms A and A' in an isotropic medium (e.g., a solution or a glass) eq. (3.53) has to be written as

$$f_{AA'}^{\mathrm{ed}} = \frac{8\pi^2 m\nu}{3he^2(2J+1)}\chi\sum_{a,a',p}(a|E_p|a')^2 \tag{3.53a}$$

If the summations are performed we obtain

$$f_{AA'}^{\mathrm{ed}} = \sum_{\lambda=2,4,6}\frac{8\pi^2 m\nu}{3h}\chi(2J+1)^{-1}\Omega_\lambda|(\mathrm{f}^N J\|U_\lambda\|\mathrm{f}^N J')|^2 \tag{3.54}$$

with

$$\Omega_\lambda = (2\lambda+1)\sum_{k,q}|B_{kq}^*|^2|Y_{k\lambda}|^2(2k+1)^{-1} \tag{3.55}$$

where in $Y_{k\lambda}$ the quantum number M_J in eq. (3.52) has to be dropped.

Equation (3.54) is quite convenient for analyzing intensities in spectra which have been taken with unpolarized light in an isotropic medium. The superscripted denotes the electric dipole nature of the radiation.

We know that in an actual calculation we encounter not pure Russell–Saunders terms, but terms that are given by the intermediate coupling approximation and are thus a sum of Russell–Saunders terms. This must be taken into account when evaluating the reduced matrix element in eq. (3.54).

For convenience we mention two other definitions for measuring intensities that are sometimes used in the literature. One is the transition line strength, given by

$$S_{AA'}^{\mathrm{ed}} = e^2\sum_{\lambda=2,4,6}\Omega_\lambda|(\mathrm{f}^N J\|U_\lambda\|\mathrm{f}^N J')|^2 \tag{3.56}$$

The other is the coefficient for spontaneous light emission from state A to state A'

$$A_{AA'}^{\mathrm{ed}} = \frac{8\nu^2 e^2\pi^2}{c^3 m}\cdot f_{AA'}^{\mathrm{ed}}\ \text{(State } A \text{ has } J) \tag{3.57}$$

The rate for spontaneous emission by electric dipole radiation is then given by using eq. (3.54)

$$A^{\text{ed}}_{AA'} = \frac{64\pi^4 e^2 \nu^3}{3hc^3(2J+1)} \chi \sum_{\lambda=2,4,6} \Omega_\lambda |(\text{f}^N J \| U_\lambda \| \text{f}^N J')|^2$$
$$A^{\text{ed}}_{AA'} = \frac{64\pi^4 \nu^3}{3hc^3(2J+1)} \chi S^{\text{ed}}_{AA'} \tag{3.58}$$

In applying these formulas to the transition between two particular J terms, we have to assume that all crystal field states are equally populated and that the transitions between two terms all have the same energy. All these assumptions are, of course, questionable but they simplify the actual calculations considerably, and if we keep in mind that the Judd–Ofelt theory involves other drastic approximations, those just mentioned are not inordinate.

The Ω_λ parameters have so far been assumed to arise solely from the crystal field; however, they also contain contributions from admixtures by the lattice vibrations. If for small elongations ΔR, the vibrational term is written as $(\partial B_{kq}/\partial R_j)\Delta R_j$, the contributions from this term are incorporated in the Ω_λ, provided the matrix elements of the vibrational coordinates are evaluated properly [Ju62].

At this point we are not yet able to decide whether the crystal field part or the vibrational part contributes most to the observed intensities. We can, however, give some qualitative arguments relating to this decision. Experimental findings show that there is indeed a coupling to the lattice vibrations because the temperature-dependent linewidths are caused by electron–phonon coupling, and transitions that involve the simultaneous excitation of an electron and a phonon (phonon sidebands) are observed. These two experimental observations demonstrate the existence of the vibrational coupling, and therefore this coupling must also mix excited configurations into the 4f^N ground state configuration. Hellwege [He41] has shown that for the RE ions, electron–phonon coupling follows a parabolic path, with the minimum of the curve at Gd^{3+}, meaning that coupling is strongest at the beginning and end of the RE series and weakest in the middle. This can be interpreted to mean that for ions like Pr^{3+}, Nd^{3+}, Er^{3+}, and Tm^{3+} the intensity has a strong vibrational coupling contribution, whereas the intensity is presumably produced mostly by the crystal field for ions like Sm^{3+}, Eu^{3+}, and Gd^{3+}.

The magnetic dipole radiation comes from the matrix element of $\mu_B(\mathbf{L} + 2\mathbf{S})$. We get the equations for magnetic dipole transitions from those of the electric dipole radiation by just replacing the operator of the electric dipole moment with that of the magnetic dipole moment. Thus for the transition between two crystal field levels a and a' we obtain (md stands for magnetic dipole)

$$f_{aa'}^{\mathrm{md}} = \frac{8\pi^2 m\nu}{he^2} \chi' |\mu_{\mathrm{B}}|^2 |(\mathrm{f}^N J M_J | L + 2S | \mathrm{f}^N J' M'_J)|^2 \tag{3.59}$$

($\chi' = n^3$, n being the refraction index.) If the summation over the three polarizations is performed, and if the average over the crystal field components is taken for the transition in an isotropic medium between two terms A and A' this is written

$$f_{AA'}^{\mathrm{md}} = \frac{8\pi^2 m\nu}{3he^2(2J+1)} \chi' |\mu_{\mathrm{B}}|^2 |(\mathrm{f}^N J \| \mathbf{L} + 2\mathbf{S} \| \mathrm{f}^N J')|^2 \tag{3.60}$$

Then we define the line strength for a magnetic dipole transition

$$S_{AA'}^{\mathrm{md}} = \mu_{\mathrm{B}}^2 |(\mathrm{f}^N J \| \mathbf{L} + 2\mathbf{S} \| \mathrm{f}^N J')|^2 \tag{3.61}$$

which yields the probability of spontaneous emission by magnetic dipole radiation

$$\begin{aligned} A_{AA'}^{\mathrm{md}} &= \frac{64\pi^4 \nu^3}{3hc^3(2J+1)} \chi' |\mu_{\mathrm{B}}|^2 |(\mathrm{f}^N J \| \mathbf{L} + 2\mathbf{S} \| \mathrm{f}^N J')|^2 \\ A_{AA'}^{\mathrm{md}} &= \frac{64\pi^4 \nu^3}{3hc^3(2J+1)} \chi' S_{AA'}^{\mathrm{md}} \end{aligned} \tag{3.62}$$

The Judd–Ofelt formalism has been applied to the analysis of a number of systems. In most of these analyses the crystal field splitting of the terms was neglected; therefore, the total absorption intensities between the ground term and the excited terms were analyzed with only three empirical parameters $\Omega_\lambda (\lambda = 2, 4, 6)$, and the success may be taken as a justification for the procedure. In a few cases, however, the intensity of transitions between individual crystal field levels were analyzed (see, e.g., [Be71]).

It is instructive to inspect the results of a typical case, such as Er in Y_2O_3, which along with other RE ions in the same host, was analyzed by Krupke [Kr66]. The oscillator strength for absorption transitions to 15 excited states was measured and then analyzed in terms of the three Ω_λ $(\lambda = 2, 4, 6)$ parameters. Comparison of the measured and calculated oscillator strengths is shown in Table 27. Except in the $^4I_{13/2}$ term, only

TABLE 27

MEASURED AND CALCULATED f NUMBERS FOR TRANSITIONS TO THE GROUND TERM $^4I_{15/2}$ IN $Er^{3+}:Y_2O_3$[a]

Initial term	Energy (cm^{-1})	f number (Meas.) (10^{-6})	f number (Calc.) (10^{-6})
$^4I_{13/2}$	6458	1.25	{0.52 md; 0.73 ed}
$^4I_{11/2}$	10,073	0.34	0.27
$^4I_{9/2}$	12,287	0.31	0.37
$^4F_{7/2}$	15,071	1.73	1.44
$^4S_{3/2}$	18,072	0.34	0.24
$^2H_{11/2}$	18,931	11.03	9.95
$^4F_{7/2}$	20,267	1.13	1.34
$^4F_{5/2}$, $^4F_{3/2}$	22,050	0.40	0.48
$^2H_{9/2}$	24,304	0.61	0.67
$^4G_{11/2}$	26,074	20.55	21.55
$^2P_{3/2}$	31,186	0.03	0.07
$^4G_{7/2}$	33,697	0.16	0.40
$^2D_{5/2}$	34,600	0.09	0.07
$^4G_{9/2}$	36,350	0.57	1.22
$^4D_{5/2}$, $^4D_{7/2}$	38,750	8.89	10.31

[a] From [Kr66].

electric dipole radiation is observed (ΔJ selection rule). Considering the crudeness of the model, it seems to be a very useful approach of data interpretation since the calculations reproduce the data quite well. The Ω_λ parameters obtained in this work for $Er^{3+}:Y_2O_3$, together with those for the other RE ions in the same host, are shown in Table 28. The Ω_λ parameters show the same kind of variation between different RE ions as was seen, e.g., for the crystal field parameters (Tables 10a and 10b).

The experimental difficulties in work like Krupke's [Kr66] are not trivial. The most important uncertainty is the concentration of RE impurities. For a host like Y_2O_3, which has two RE sites, one with point symmetry C_2 (measured) and one with point symmetry C_{3i} (no allowed electric dipole radiation because of inversion center), an additional source of possible error comes from the uncertainty of the occupation of the two sites.

TABLE 28

COMPILATION OF JUDD–OFELT INTENSITY PARAMETERS Ω_λ FOR RARE EARTHS IN Y_2O_3[a]

System	Ω_2 (10^{-20} cm^2)	Ω_4 (10^{-20} cm^2)	Ω_6 (10^{-20} cm^2)
Pr^{3+}:Y_2O_3	17.21	19.8	4.88
Nd^{3+}:Y_2O_3	8.55	5.25	2.89
Eu^{3+}:Y_2O_3	6.31	0.66	<0.48
Er^{3+}:Y_2O_3	4.59	1.21	0.48
Tm^{3+}:Y_2O_3	4.07	1.46	0.61
Yb^{3+}:Y_2O_3[b]	3.5	1.75	0.80

[a] From [Kr66].

[b] The values listed for Yb^{3+} : Y_2O_3 were determined by extrapolation of the experimental parameters of the other REs.

So far, attempts to calculate Ω_λ parameters from first principles have been unsuccessful. These calculations were hampered by many factors: They started from a point charge model (which we know is a bad approximation) and had to use approximate values for the radial integrals involving the excited configurations. Table 29 contains a number of Ω_λ parameters obtained in different hosts for one RE ion, Nd^{3+}. We see that the Ω_2 parameters vary greatly among different hosts, whereas the Ω_4 and Ω_6 parameters are quite constant. This means we can use extrapolated parameters for the latter two to predict intensities in crystals in which they have not yet been measured.

TABLE 29

JUDD–OFELT INTENSITY PARAMETERS Ω_λ FOR COMPOUNDS OF Nd^{3+}[a]

Ω_λ ($\times 10^{20}$)cm^2	λ: 2	4	6	Point symmetry	Reference
Nd^{3+} aquo	0.9	5.0	7.9		[CaFi68a]
$NdBr_3$	180	9	9	D_{3h}	[GrKo67]
NdI_3	275	9	9	D_{3h}	[GrKo67]
Nd^{3+}:YAlG	0.2	2.7	5.0	D_2	[Kr71]
Nd^{3+}:Y_2O_3	8.6	5.3	2.9	C_2	[Kr66]
Nd^{3+}:LaF_3	0.4	2.6	2.5	C_2	[Kr66]
Nd^{3+}:$YAlO_3$	1.3	4.7	5.7	C_{1h}	[WeVa71, 73]

[a] From [Pe75].

Much of the research work on intensities has been done in solutions [CaFi68a], which have the advantages that the crystal field splitting in them is mostly wiped out right away (because of the large linewidths produced by the noncrystalline environment and the high temperature of 300 K), and that it is quite easy to determine the RE concentrations. A good compilation of available experimental data is found in the work of Peacock [Pe75] and Reisfeld [Re73, Re75].

To conclude this section, we comment briefly on the so-called hypersensitive transitions [JoJu64, Ju66b]. These are transitions where the intensity is particularly sensitive to the host; typical examples are $^3H_4 \rightarrow {}^3F_2$ (Pr^{3+}), $^6H_{15/2} \rightarrow {}^7F_{11/2}$ (Dy^{3+}), and $^3H_6 \rightarrow {}^3F_4$ (Tm^{3+}). (For a compilation see [Pe75].) With possible exceptions, these transitions follow the selection rules $\Delta J \leq 2$, $\Delta L \leq 2$, and $\Delta S = 0$, meaning that their intensity is solely determined by the Ω_2 term [Pe75, Bl76]. Thus to explain the phenomenon we require mechanisms that strongly influence Ω_2, while leaving the two other terms largely unaltered. A number of such mechanisms [Pe75] have been considered, such as the A_{1q} term in the crystal field [Ju66b], lattice vibrations [JoJu64], covalency [HeCh68], quadrupole radiation [JoJu64], inhomogeneous dielectric (variation of the electric field across the ion) [JoJu64], and a dynamic coupling (mixing of the RE ligand wavefunctions due to inclusion of unoccupied orbitals) [Pe75] [PcSte78], but so far no consistent explanation of hypersensitivity has been found. It is probably caused by a combination of some of the mechanisms mentioned.

3.16 Phonon Sidebands

In the discussion of the number of lines in the spectra we have assumed thus far that the RE ion rests in an ideal crystal and there is no coupling to either other RE ions or the lattice vibrations. A look at typical low temperature optical RE absorption spectra shows that this cannot be true: Invariably, more lines are observed for a transition from the crystal field ground state to an excited crystal field multiplet than are expected from Table 22. This observation raises two questions: How can actual spectra nevertheless be used to determine, e.g., the J value of an excited term (in other words, how can we sort out from the many lines observed those that "truly" correspond to transitions between individual crystal field levels), and what are the mechanisms that give rise to the many "extra" lines?

The first questions cannot be answered exactly; we can offer only a few guidelines based on experimentalist's experience. As a rule of thumb we can assume that the "true" crystal field transitions are the strongest lines in a spectrum. This means that in most cases it is advisable when analyzing a spectrum to dilute the concentration of RE ions by diamagnetic analogs (La, Lu, Y) and to note which lines survive this procedure. In cases with high symmetry and correspondingly few crystal field parameters, it may also be advisable to determine the crystal field parameters from crystal field multiplets, where analysis is possible, and then calculate the expected crystal field pattern. This method is often useful for eliminating "impossible" assignments. We have tacitly assumed here that the J value of the excited term is known, e.g., from a comparison of the energy of the excited level with the diagram in Fig. 5.

With respect to the extra lines, they originate from three main sources: the electron phonon coupling (see Chapter 4), which causes joint transitions in the electron–phonon systems; impurities in the crystal ("satellites"); and magnetic interactions (which will be dealt with in Chapter 6). This section will be devoted to phonon sidebands; the next section will briefly treat satellite lines.

The lattice ions are not fixed; they vibrate, and because the lattice potential is of necessity anharmonic, there is a coupling of the 4f electrons to the lattice vibrations. Various aspects of this coupling will be dealt with in the next chapter—here we are interested only in the fact that via this coupling, joint transitions in the electron–phonon system are possible. Thus, every electronic line is in principle accompanied, even at $T = 0$, by a number of lines that originate from the simultaneous excitation of phonons (selection rules permitting, of course). Although some research has been done in this area, in practice little use can be made of RE spectra in deducing information on the phonon system. Much more powerful methods, like neutron scattering, Raman scattering, and infrared absorption measurements, are called for. Thus in most cases phonon sidebands have been considered rather an unwanted complication than an actual source of information.

Due to the weak interaction of the 4f electrons with the lattice and its vibrations, a description within the Born–Oppenheimer approximation, which assumes that a separation of the nuclear and electronic coordinates is possible, is appropriate. The wavefunction of the total system Ψ is then the product of an electronic part Ψ_e and a vibrational part Ψ_v.

$$\Psi = \Psi_e \cdot \Psi_v$$

In this approximation the transition probability w is given as a product of an electronic factor and a lattice factor [ImYe64]

$$w \sim |\langle \Psi_{e1}|E|\Psi_{e2}\rangle|^2 \cdot \frac{D^*(\omega)}{\omega^3} \tag{3.63}$$

where E is the operator of the electric dipole moment, 1 and 2 are the two electronic states between which the optical transition takes place, and $D^*(\omega)$ is an effective phonon density of states. (Effective means that it is weighted by selection rules and by a frequency dependent coupling constant.) The simplified eq. (3.63) shows the essential physics: The pure electronic transition is accompanied by a phonon sideband that (though modified) reflects the phonon density of states; we keep in mind that the most severe restriction comes from the selection rules. If we assume complete separation of electronic and vibronic wavefunctions, only transitions between phonon states of the same symmetry are possible.

Although at first sight this method seems an appealing one by which to determine phonon density of states, it has actually been used only to prove the correctness of the foregoing description rather than to obtain information on the phonons. The vibronic spectrum in the RE trichlorides has been investigated in considerable detail [CoMo67a, CoMo67b, RiSa63, Sa64, CoRi68]; however, since their structure is relatively complicated, only limited comparison with other data could be made.

More favorable in this respect are simple crystals like the alkali halides, about whose phonon spectra detailed neutron data exist. However, RE ions can only be incorporated into alkali halides as divalent impurities; thus a charge compensation center is produced that reduces the site symmetry for the RE ion.

Here we have to distinguish two cases. For the divalent RE ions the $4f^N \rightarrow 4f^{N-1}5d$ transitions (which are parity allowed) fall into the visible region, and can be seen along with the $4f^N \rightarrow 4f^{N'}$ transitions. The $4f^N \rightarrow 4f^{N-1}5d$ transitions couple strongly to the lattice (due to the extended nature of the 5d wavefunction) and therefore produce a vibronic structure distinctly different from that of the weak $4f^N \rightarrow 4f^{N'}$ transitions; we shall not discuss that structure here (see [Wa65, WoKa62, AxSo63, WaBr65a, WaBr65b]).

The $4f^N \rightarrow 4f^{N'}$ transitions of RE ions in alkali halides, however, show a weak vibronic structure that falls into the categories discussed above. The emission spectrum of the Sm^{2+}:KBr system has been investigated in some detail because it reveals the favorable ${}^5D_0 \rightarrow {}^7F_0$ [Br65, TiBu67] transition, which consists of a single line well separated from other lines, thus permitting clear discernment of the complete vibronic sideband. In principle, this transition is forbidden in the cubic KBr structure; however, the charge compensating center destroys the cubic symmetry, allowing this transition. Timusk and Buchanan [TiBu67] compared the measured vibronic sideband and the one calculated from the dispersion relations for KBr [CoCo63] obtained by analyzing the experimental dispersion

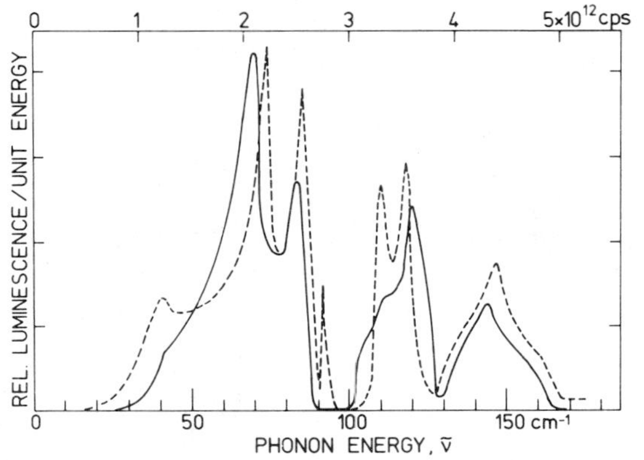

Fig. 14. Comparison of the measured phonon sideband intensity in the emission spectrum of the $^5D_0 \rightarrow {}^7F_0$ transition in Sm^{2+} : KBr [TiBu67] (dashed line) and the phonon density of states [CoCo63] of KBr as determined from neutron scattering experiments (solid line).

The zero phonon line ($^5D_0 \rightarrow {}^7F_0$) is positioned at the zero of the energy scale and not shown.

relations with the shell model [DiOv58]. Figure 14 compares the calculated with the observed vibronic sidebands, showing quite good agreement. The most serious difference, the consistently lower energy of the theoretical spectrum, may be caused by the fact that data taken at 90 K were used in the theory, whereas the optical experiment was performed at 7 K.

In more complicated crystals, the optical phonons and, where they exist, the internal vibrations of complexes forming the crystal can give rise to a rich sideband structure. A nice example is the spectrum of $Pr_2Zn_3(NO_3)_{12} \cdot 24H_2O$ (praseodymium zinc double nitrate), which shows up to about 40 sideband lines accompanying a single electronic transition [HeHe52]. These many lines considerably complicates the analysis of a spectrum.

3.17 Satellites

The phonon sidebands discussed in the preceding section have their main intensity, usually 50 cm^{-1} and more, away from the electronic line. However, if we look closely at a typical absorption spectrum of RE compound (e.g., $PrCl_3$), we see that in the immediate vicinity of a crystal field transition there are a great number of other lines, called satellites [HeHe51, PrCo68]. These satellites almost invariably show the same polarization behavior as the main line, and in high magnetic fields they have the same Zeeman pattern. The great number of satellites observed makes it unlikely that they have a common cause. Faulhaber [Fa69] and Yamamoto *et al.* [YaOt69] produced satellites in RE crystal spectra and in both experiments the lines were the results of impurities. Yamamoto *et al.* [YaOt69] doped crystals with other RE impurities which have slightly different ionic radii than the original RE ions. Thus, the RE impurities produce a local distortion and thereby a different crystal field, which leads to different energy levels and additional lines in the absorption spectra. Although this is in principle a source of satellites, the work in this field has shown that it is probably not the most frequent one. In addition, it is easy to avoid these satellites, because very pure RE samples are available commercially these days.

Faulhaber [Fa69] doped crystals of DAG (dysprosium aluminium garnet, $3\,Dy_2O_3 \cdot 5\,Al_2O_3$) with small amounts of gallium, thereby producing satellites. This experiment demonstrated that if in a crystal a regular non-RE ion is replaced by one differing in radius and ionicity, additional lines are produced. Further experiments on other crystals [Do71, BuDo71, He72, HeHe72a, HeHe72b, He73, HeHe73, BaPe75] demonstrated that satellites could be produced by this method, indicating that many of the observed satellites probably had their origin in such non-RE impurities. The deliberately introduced impurities constitute only a minor perturbation of the original lattice, making it understandable that the properties of the satellite lines are very similar to those of the main crystal field lines. So far we can distinguish three mechanisms by which impurities produce shifts of the energy levels in a crystal with respect to those of the unperturbed crystal. The impurity can change the symmetry of the environment of an RE ion [Bu69]; it can have a different ionicity [Fa69] from the ion it has replaced; and it can have a different ionic volume [BaPe75].

The effects of a change in symmetry can most easily be seen in systems with an even number of electrons; here in a crystal of relatively high symmetry a doubly degenerate level can be split into two singly degenerate levels. This was demonstrated by Heischmann *et al.* [HeHe72a,b] who doped praseodymium double nitrate crystals ($(Pr_2Zn_{3c}Mg_{3(1-c)}(NO_3))_{12}\cdot 24H_2O$) with small amounts of Mn, thus destroying the C_{3v} symmetry at the site of the RE ion. The doubly degenerate ground state level in the pure crystal was then split by the crystal field perturbation. This can be seen in analyzing the $^3H_4 \rightarrow {}^3P_0$ transition (the $^3P_0$ term is not influenced by the crystal field in first order), which gives the splitting pattern of the ground state. Figure 15 shows the magnetic field dependence of this transition in the doped crystal. The main lines (M_+, M_-) show a linear Zeeman effect (as in the undoped crystal) and the two satellite lines (σ_1, σ_2) show a quadratic Zeeman effect (σ_+, σ_-). This is easily understood, because in the satellite energy level scheme there are, in addition to the Zeeman matrix element, those of the crystal field perturbation ΔV. If the shift of the total energy field perturbation is ΔV_0, and the splitting of the degenerate level by the crystal field perturbation is ΔV_1, we obtain for the Zeeman pattern

$$W_{\sigma+/\sigma-} = \Delta V_0 \pm ((\Delta V_1/2)^2 + (s\mu_B H)^2)^{1/2} \tag{3.64}$$

where the subscript σ_+/σ_- designates the two satellites and s is the magnetic splitting factor of the unperturbed energy level. We see in Fig. 15 that this describes the experimental results well, although not perfectly.

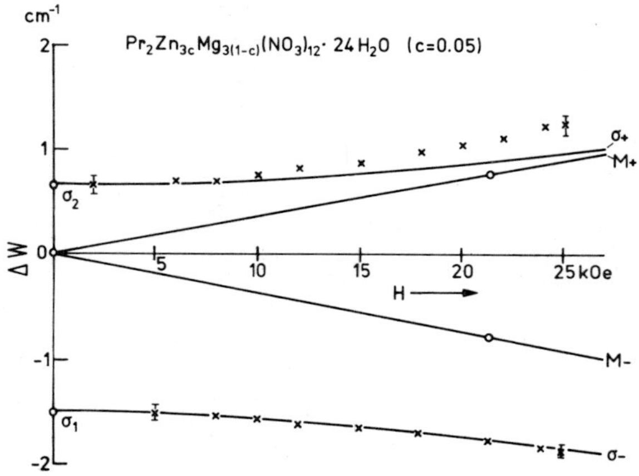

Fig. 15. Magnetic field dependence of the $^3H_4 \rightarrow {}^3P_0$ transition in $Pr_2Zn_{3c}Mg_{3(1-c)}(NO_3)_{12}$[HeHe72a] at 4.2 K.

M_+ and M_- designate the main lines for positive and negative splitting factors, σ_+ and σ_- the same lines for the satellites. The line connecting the satellite data points is calculated with eq. (3.64), the main lines show a linear Zeeman effect.

Baumert *et al.* [BaPe75] investigated satellites in Nd^{3+}:$LaCl_3$ doped with Br and in Nd^{3+}:LaBr doped with Cl. This system does not readily show the effect of a lowering of the symmetry because all the crystal field levels are Kramers doublets and will stay such in a lower symmetry crystal field; but it reveals an effect of the different sizes of the Cl and Br ions and of their different ionicity. Upon doping with Br, the $^4I_{9/2} \rightarrow {}^4S_{3/2}$ transition in Nd^{3+}:$LaCl_3$ shows up to nine satellites (corresponding to nine halogen ions in the immediate environment), with an average separation of $\delta = 7.9\ cm^{-1}$. The satellites occur at lower energy with respect to the main lines.

In addition to the satellite lines, we observe a shift of the center of gravity of the main lines of the $^4I_{9/2} \rightarrow {}^4S_{3/2}$ transition to higher energies. This shift can be interpreted as due to a change in the atomic volume and is to higher energies with increasing volume, because the $LaBr_3$ lattice has a larger volume compared to the $LaCl_3$ lattice (the sign of the shift is consistent with the results obtained from high pressure experiments [Ba75]). If we extrapolate the change in position of the $^4I_{9/2} \rightarrow {}^4S_{3/2}$ transition by the addition of Br to 100% Br, we obtain the position of this transition in a fictitious $LaCl_3$ lattice with the $LaBr_3$ lattice parameters. In that lattice the energy of the $^4I_{9/2} \rightarrow {}^4S_{3/2}$ transition is shifted by 18 cm^{-1} with respect to its position in the $LaCl_3$ lattice. To check this result we can dope the $LaBr_3$ lattice with Cl ions, observe the change in the center of gravity of the $^4I_{9/2} \rightarrow {}^4S_{3/2}$ transition, extrapolate it to 100% Cl ions, and obtain the same number for the shift.

From this we realize that introducing Br into the $LaCl_3$ lattice (or Cl into the $LaBr_3$ lattice) entails two contributions, one from a change in the ionic volume and one from a change of the ionicity, and the data are completely consistent. Let us assume that the total energy shift due to the change in ionicity in going from Cl to Br is produced by the nine ligands. Since the energy shift produced by replacing one ligand is 7.9 cm^{-1}, a total shift of 71.1 cm^{-1} is produced. This is the energy difference between the $^4I_{9/2} \rightarrow {}^4S_{3/2}$ transition in $LaBr_3$ and in $LaCl_3$ with the lattice constant of $LaBr_3$, because only then is there no contribution to the energy shift due to the change in ionic volume. We know that the energy of the transition in $LaCl_3$ with the $LaBr_3$ lattice constant is 18 cm^{-1} larger than in pure $LaCl_3$, and the energy separation of the transition between $LaCl_3$ and $LaBr_3$ is 54 cm^{-1}, giving a total of 72 cm^{-1}, in very good agreement with the 71.1 cm^{-1} estimated from the satellite separation. For the $^4I_{9/2} \rightarrow {}^2P_{1/2}$ transition the numbers are different but they show the same consistency: The additional shift due to the volume change is 39 cm^{-1}, making

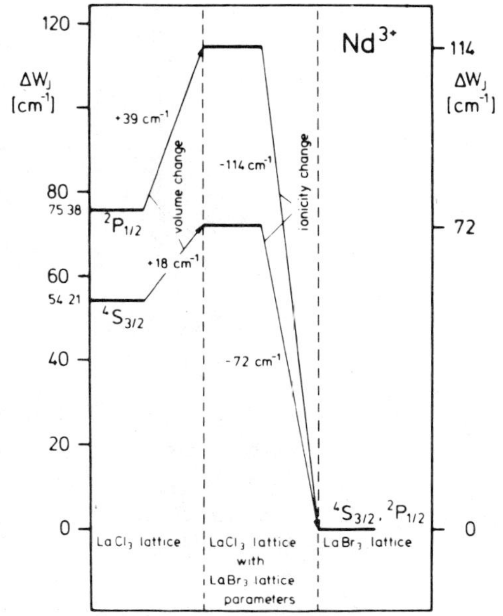

Fig. 16. Energy shift of the $^4I_{9/2} \rightarrow {}^4S_{3/2}$ and $^2P_{1/2}$ transitions in Nd^{3+} : $LaCl_3$ relative to their position in $LaBr_3$.

The experimental data are analyzed in terms of a volume shift and the nine satellite shifts (9δ) [BaPe75]. The figure shows that $\delta(^4S_{3/2}) = -72/9 = -8\ \text{cm}^{-1}$ and $\delta(^2P_{1/2}) = -114/9 = -12.7\ \text{cm}^{-1}$.

the $LaBr_3$ to $LaCl_3$ with the $LaBr_3$ lattice constant separation 114 cm^{-1}. The average satellite separation for this transition is 12.9 cm^{-1}, giving $9 \cdot 12.9 = 116.1\ \text{cm}^{-1}$, again in very good agreement with the expected value. These data are illustrated in Fig. 16, where the positions of the $^4I_{9/2} \rightarrow {}^4S_{3/2}$ transition and the $^4I_{9/2} \rightarrow {}^2P_{1/2}$ transition in $Nd^{3+}LaBr_3$ have been normalized to zero. The other energies have been drawn relative to that zero, as discussed in the text.

The experimental material in this field is too limited to permit us to draw any conclusion about the exact mechanism of the ionicity change, but we hope these satellite effects will be explored further.

CHAPTER 4

Trivalent Rare Earth Ions in a Phonon Field

4.1 Introduction

So far the crystal has been considered static,† and the effect of this static crystalline environment of 4f electrons has been called the crystal field interaction, removing the M_J degeneracy of the free ion energy levels. In reality, at finite temperatures the ions in a crystal move around their equilibrium positions, and the collective modes of these motions are the lattice vibrations. There are two types of lattice vibrations: acoustical vibrations, which start at $\omega = 0$ for $k = 0$; and optical vibrations, which have finite energies at $k = 0$, where $k = 2\pi/\lambda$ is the wave vector. Both types of vibrations have distinct effects on the spectra of REs. These effects will first be described qualitatively, and then we shall deal with them more quantitatively. (The original work on the effects of phonons on narrow optical lines in solids can be found in the papers of McCumber *et al.* [McCuStu63, ImYe64].)

The effect of the acoustical vibrations can be visualized most easily by recalling certain phenomena from the field of paramagnetic relaxation [Je63, MaOr66]. The moving ions modulate the crystal field, and this modulation can induce transitions between different crystal field levels.

† This statement does not hold for that part of the intensities that is produced by the lattice vibrations (Section 3.15) or for the phonon sidebands (Section 3.16).

Conservation of energy is satisfied by the emission or absorption of a phonon. Paramagnetic relaxation is the result of phonon induced transitions between the two components of a magnetic field split crystal field doublet, mostly the ground doublet. But the kind of transition found for this case also occurs in the more general situation of two levels split by the crystal field interaction.

Transitions between two levels can be brought about in various ways. The simplest method is the direct process. Here an ion makes a direct transition from one level to another and either the difference in energy is used to create a phonon of the same energy, or a phonon of that energy is absorbed in order to provide the energy necessary to make the transition energetically possible. Another type of transition is made possible by the inelastic scattering of phonons, or Raman process. Here a phonon is scattered from an energy level, its energy is diminished or increased in that scattering process, and this change in energy allows the transition into another energy level. Finally, there is the Orbach [Or61, FiOr61], or resonance Raman, process, which can be viewed as the consecutive action of two direct processes: Via the absorption of phonon, a crystal field state goes into an excited crystal field level, and via the emission of still another phonon, a lower crystal field level is reached. Since this process involves two direct processes, we shall not treat it as a separate mechanism, but shall consider its effect when dealing with the direct processes.

All these processes have characteristic temperature dependences, by which they can be distinguished. For our purpose it is important to note that interaction with the acoustical phonon field induces transitions between various crystal field levels, and these transitions shorten the lifetimes of the levels, thus contributing to the broadening of the spectral lines. In addition, the lattice vibrations also contribute to the crystal field energy a small temperature-dependent term that shows up in absorption spectra as a temperature-dependent line shift.

In Chapter 3 it was shown that transitions in the RE spectra are mainly electric dipole in nature and are relatively weak because they come about only through admixtures of other electronic configurations to the 4f configuration. This means that the radiative lifetime of an excited state is usually of the order of milliseconds. The measured lifetimes, which can be of the order of microseconds, are produced by radiationless transitions, which contribute the most to the decay probability of the excited states. The phonon decay processes within a given multiplet, which were discussed qualitatively earlier, give much shorter transition rates of the order of 10^{-10} sec or less. It is therefore often possible to regard the lowest crystal field level of a crystal field multiplet as the ground state and to neglect the decay to other crystal field multiplets when considering the linewidth within the crystal field multiplet itself.

In most cases, optical phonons have only a limited influence on the linewidth within a specific crystal field multiplet. On the other hand, they are responsible for the radiationless decay of the excited states. Energy considerations indicate that many optical phonons must be involved in these processes. It is believed that these many phonon processes are the main source of the depopulation probability for the optical excited states.

The optical phonons coming from, e.g., the H_2O complex of a hydrated crystal are numerous and have short lifetimes (large linewidth in the spectrum). Hence, even in the case of inexact energy matching, these phonons may be significantly responsible for these depopulation processes, because the large wings of the energy distributions of the phonons often conduce to an appreciably overlap and thus to a large transition rate. Therefore, hydrated crystals show far fewer intense radiative transitions than anhydrous crystals.

We shall not discuss the effect of impurities and defects on the lifetimes of the crystal field levels. The following description will show that for the linewidth within a crystal field multiplet, scattering via defects is probably small. On the other hand, it is hard to estimate the effect of impurities on the radiative lifetime of an excited state. Very few data on this relationship exist, and since they are not very conclusive, we shall neglect them.

Finally, the magnitude of the coupling of the active ion to the phonon field is basically very important for the occurrence or nonoccurrence of the sharp optical lines in spectra of solids. The weak electron–phonon coupling for the RE only makes the occurrence of their optical lines in solids possible. The underlying physics will be treated briefly in what follows.

4.2 Optical Analog to the Mössbauer Effect

Soon after the discovery of the Mössbauer effect [Mö58] it was realized that there is a strong analogy between the occurrence of the narrow γ line in this case and the narrow optical lines seen in the spectra of 3d ions and RE ions in solids, as well as the lines produced by color centers [SiFi64].

Mössbauer's discovery [Mö58] consisted in his finding that if radioactive nuclei imbedded in a solid emit a low energy γ ray ($W_\gamma < 150$ keV), a certain amount of these γ-ray decays (the so-called recoil free fraction) occurs in such a way that the energy of the γ ray is the full γ energy and more important the width of this line is determined by the lifetime of the decaying excited nuclear state. This is not a trivial observation because the decay produces a recoil energy $\hbar^2k^2/2m$ (where $\hbar k$ is the momentum of the emitted γ ray and m the mass of the atom) that if taken up by the emitting atom alone and transmitted into the phonon system, would produce a considerable energy shift and a broadening of the γ line.

In a classical picture the Mössbauer effect can now be thought of as a modulation of the γ-ray frequency by the lattice vibrations. The central line (carrier frequency), which has approximately the full γ-ray energy, is called the phonon-free transition because in this case the recoil produced by the emission of the γ ray is taken up by the crystal as a whole and thus produces only a negligible energy shift (m is replaced by M, the mass of the total crystal). The possibility remains, of course, that the recoil is taken up by the phonon system, in which case there is a large energy shift due to the excitation of the phonons. The phonon-free fraction of γ rays emitted by a given crystal is given (for $T = 0$ and assuming a Debye spectrum for the lattice vibrations) [Ma66] by

$$f = \exp(-3W_r/2k_B\Theta), \tag{4.1}$$

where W_r is the recoil energy given by the emitted γ ray to the emitting atom (often written as $\hbar^2k^2/2m$), k_B is Boltzmann's constant, and Θ the Debye temperature. We see that for a given recoil energy the recoil free fraction gets larger with increasing Θ or, for a given crystal structure, with increasing strength of the crystal forces: the stronger the crystal forces, the larger the recoil-free fraction.

This phenomenon has a strong formal resemblance to the narrow optical lines seen in RE spectra (as well as in spectra of 3d compounds and especially for the narrow zero phonon lines† seen in the spectra of color centers in alkali halides [FiSi63, Tr63, SiFi64]). Electron–phonon coupling leads, as shown in Section 3.17, to electron–phonon transitions (phonon sidebands), in which are seen not the pure transition between electronic energy levels, but one in which, in addition, one or more phonons are excited. It is now evident that the stronger the electron–phonon coupling the greater the probability for transitions in which a phonon is also created relative to the probability for zero phonon transitions. Thus in the case of these optical lines the electron–phonon coupling strength plays a role similar to

† By the term zero phonon line an optical line is meant that originates from a pure electronic transition; the phonon system is the same in the initial and the final state, thus, e.g., all the lines except the phonon sidebands of the RE spectra that we deal with in this book are zero phonon lines.

that of the recoil energy in the case of the Mössbauer effect. Hence there emerges from the rigorous theory [Ma66] a factor governing the intensity of the zero phonon line relative to all transitions; it has the form (in the Debye approximation and for $T = 0$)

$$f_1 = \exp\{-(9n/2)(W_d/k_B\Theta\} \tag{4.2}$$

where n is the number of atoms in the unit cell. W_d is the difference in energies for the two energy levels between which the optical transition takes place, calculated without electron–phonon coupling and by taking the linear term in the electron–phonon coupling (see Section 4.3) to second order perturbation theory. It is thus a measure of the electron–phonon coupling strength.

This factor f_1 has the same form as the factor f determining the recoil-free fraction in the Mössbauer spectra; it gives the fraction of transitions that are observed without the simultaneous excitation of a phonon. In this case the place of the recoil energy is taken by the electron–phonon coupling strength. It is verified that for two limiting cases the f_1 factor gives the correct result. For zero electron–phonon coupling, $f_1 = 1$ and, as expected, all transitions are phonon free; for very large values of W_d the value of f_1 gets very small, meaning that the phonon-free transition is not observed at all. As mentioned before, due to the inner shell nature of the 4f orbitals, the electron–phonon interaction is small (leading to W_d values of the order of 1 cm^{-1}; the value of $k_B\Theta$ is of the order of 100 cm^{-1}) and the fraction of phonon-free transitions is therefore close to 1. On the other hand, in the case of the lines produced by defects (e.g., in alkali halides) the electron–phonon coupling can be very large, and there may be no phonon-free transition at all.

4.3 Effects of Acoustical Phonons

The interaction of the magnetic moment of an ion with the phonon field in an insulator has been dealt with by many authors attempting to explain paramagnetic relaxation phenomena [Or61, Ste67, ScJe62]. This approach was also applied to the interpretation of optical linewidths and line shifts for similar systems and we shall follow these authors [YeSc64]. The relevant Hamiltonian contains the terms

$$\mathscr{H} = \mathscr{H}_I + \mathscr{H}_P + \mathscr{H}_{IP} \tag{4.3}$$

where $\mathscr{H}_I$ stands for the energy of the paramagnetic ion (free ion plus static crystal field energy), $\mathscr{H}_P$ stands for the phonon field energy, which we assume to be an ensemble of harmonic oscillators; and $\mathscr{H}_{\text{IP}}$ is the term that represents the interaction between the two systems. This last term is the important one because it is responsible for the transitions between various crystal field levels by emission and absorption of phonons. Our exposition of the problem derives from Orbach [Or61], who suggested developing the crystal potential in a power series of the strain. The energy level scheme relevant to this discussion appears in Fig. 17, which shows a typical crystal field multiplet where the total splitting is of the order of $k\Theta$. This means that the energy difference between the various crystal field levels $|a_i)$ is of the order of an acoustical phonon. The static crystal field Hamiltonian is

$$\mathscr{H}_{\text{CF}} = V = \sum_{kq} A_{kq} P_{kq}$$

We now assume that the Hamiltonian for a position $R + \Delta R$, where R is the equilibrium position, can be obtained by a series expansion

$$V(R + \Delta R) = V(R) + \frac{\partial V}{\partial R}\Delta R + \frac{1}{2}\frac{\partial^2 V}{\partial R^2}(\Delta R)^2 + \cdots \tag{4.4}$$

If we introduce a homogeneous lattice strain $\varepsilon = \Delta R/R$, this expansion can be written as [Or61]

$$V(R + \Delta R) = V_0 + V_1\varepsilon + \tfrac{1}{2}V_2\varepsilon^2 + \cdots \tag{4.5}$$

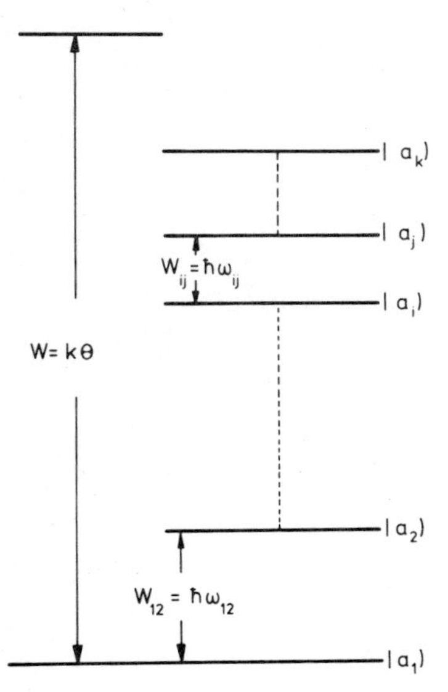

Fig. 17. Energy level diagram for a discussion of the line broadening within a crystal field multiplet [YeSc64]. θ is the Debye temperature.

Using the Hamiltonian for the crystal field interaction we obtain

$$V = V_0 + \frac{\partial}{\partial R}\left(\sum_{k,q} A_{kq}\right) \Delta R\, P_{kq} + \frac{1}{2}\frac{\partial^2}{\partial R\,\partial R'}\left(\sum_{kq} A_{kq}\right) \Delta R\, \Delta R'\, P_{kq} + \cdots \tag{4.6a}$$

and by using the strain ε we get

$$V = V_0 + \varepsilon R \frac{\partial}{\partial R}\left(\sum_{kq} A_{kq}\right) P_{kq} + \tfrac{1}{2}(\varepsilon\varepsilon' RR') \frac{\partial^2}{\partial R\,\partial R'}\left(\sum_{kq} A_{kq}\right) P_{kq} + \cdots \tag{4.6b}$$

If we assume that $A_{kq} \sim 1/R^n$, we obtain as an order of magnitude estimate

$$V_{\mathrm{IP}} = \varepsilon \sum_{k,q} A_{kq} P_{kq} + \tfrac{1}{2}\varepsilon^2 \sum_{kq} A_{kq} P_{kq} + \cdots \tag{4.6c}$$

which means that in the first approximation the dynamical crystal field is just the strain times the static crystal field.

Although this approach is widely used to describe the electron–phonon interaction, a few remarks about its validity are in order. First, the crystal field parameters $A_{kq}\langle r^k \rangle$ depend not only on R, the distance of the ligands, but also on the ligands' angular positions. If the lattice vibrations alter the positions of the ligand, we have to calculate these derivations as well, and include these terms in the electron–phonon interaction. The differentiation with respect to R gives a factor n, the power of $1/R$. Newman's work [Ne71] shows that n is somewhere between 5 and 10, indicating that this approach gives numerical results that may be inaccurate by as much as an order of magnitude. The approach outlined, therefore, sacrifices accuracy to ease of handling.

A transition from a state a_i to a state a_j by means of the electron–phonon interaction accompanied by the emission of a phonon (i.e., by the direct process) is then calculated as

$$w^{\mathrm{d}}_{i\to j} = \frac{2\pi}{\hbar} |\langle i| V_{\mathrm{IP}} |j\rangle|^2 D(W) \tag{4.7a}$$

where $D(W)$ is the density of final states. Using V_{IP} in eq. (4.6c) to first order yields if $W_i > W_j$

$$w^{\mathrm{d}}_{i\to j} = \frac{2\pi}{\hbar} D(W) |\langle p(W_{ij})|\varepsilon| p(W_{ij})+1\rangle|^2 |\langle i| \sum_{kq} A_{kq} P_{kq} |j\rangle|^2 \tag{4.7b}$$

where $p(W_{ij}) = [\exp(W_{ij}/kT) - 1]^{-1}$ is the average phonon excitation number. In the Debye approximation the phonon density of states is given by [Ki71]

$$D(W)dW = 3VW^2\, dW/2\pi^2\hbar^3 v^3 \tag{4.8}$$

with V the volume of the crystal and v the average velocity of sound. The square of the matrix element of the strain is $W_{ij}[p(W_{ij}) + 1]/2Mv^2$, where M is the mass of the crystal [AbBl70]. So we obtain for the transition rate

$$w_{i\to j}^{\mathrm{d}} = \frac{3}{2\pi\rho v^5\hbar}\left(\frac{W_{ij}}{\hbar}\right)^3 |\langle i|\sum A_{kq}P_{kq}|j\rangle|^2[p(W_{ij}) + 1] \tag{4.7c}$$

where $\rho = M/V$ is the density of the crystal.

Since we shall be interested only in the temperature dependence of the transition rate, we write

$$w_{i\to j}^{\mathrm{d}} = K_{ij}[p(W_{ij}) + 1] \qquad \text{for} \quad j < i \tag{4.7d}$$

(where the one in the brackets takes care of the spontaneous emission rate) and

$$w_{i\to j}^{\mathrm{d}} = K_{ij}p(W_{ij}) \quad \text{for} \quad j > i \tag{4.7e}$$

The total transition rate from the crystal field level i to all other crystal field states j of the same crystal field multiplet due to direct emission or absorption of phonons is given by

$$w_i^{\mathrm{d}} = \sum_{j<i} K_{ij}[p(W_{ij}) + 1] + \sum_{j>i} K_{ij}p(W_{ij}) \tag{4.9}$$

Another process that can produce transitions from crystal field level i to another crystal field level j is the Raman process, or the inelastic scattering of phonons. This process is calculated by taking in eq. (4.6c) the term linear in ε to second order or by taking the term in ε^2 to first order, which (after setting $x = W/k_{\mathrm{B}}T$) yields

$$\begin{aligned} w_i^{\mathrm{R}} &= H_i\left(\frac{T}{\Theta}\right)^7 \int_0^{\Theta/T} dx \, \frac{x^6 e^x}{(e^x - 1)^2} \\ &= H_i\left(\frac{T}{\Theta}\right)^7 I_6\left(\frac{\Theta}{T}\right) \end{aligned} \tag{4.10}$$

where $I_6(\Theta/T)$ is an integral tabulated by Ziman [Zi54] and H_i a combination of crystal field matrix elements (including some constants) whose exact form does not interest us, because again we are interested only in the temperature dependence of the transition rate (see, e.g., [Or61, YeSc64, Ku69]).

In some cases the energy separation between two crystal field levels is larger than the Debye energy $k\Theta$. Then the crystal field level can decay only via the spontaneous emission of several phonons. This decay rate is temperature independent and relatively small

$$w_i^s = \sum_{j<i} \alpha_{ij} \tag{4.11}$$

The total linewidth $\Delta\tilde{\nu}$ (in cm^{-1}) of a crystal field level i is then given as the sum of the contributions mentioned earlier:

$$\Delta\tilde{\nu} = \frac{1}{2\pi c}\sum_{j<i}\alpha_{ij} + \frac{1}{2\pi c}\sum_{j<i}K_{ij}[p(W_{ij}) + 1] + \frac{1}{2\pi c}\sum_{j>i}K_{ij}p(W_{ij}) + \frac{1}{2\pi c}H_i\left(\frac{T}{\Theta}\right)^7 I_6\left(\frac{\Theta}{T}\right) \tag{4.12}$$

The width of an optical absorption or emission line is always the sum of the linewidth of the two levels o and p between which the actual transition takes place.

$$\Delta\tilde{\nu}_{o\to p} = \Delta\tilde{\nu}_o + \Delta\tilde{\nu}_p \tag{4.13}$$

The mechanisms described so far yield a homogeneous broadened optical line, meaning a Lorentzian line shape.

Random strains in the crystal produce an inhomogeneous contribution to the linewidth, generally yielding a Gaussian line shape (but see [Sto69]). This contribution has to be subtracted from the measured linewidth in order to derive the true lifetime broadening.

There have been a great many investigations of linewidths and their temperature dependence. We shall discuss a representative one, using the data of Yen *et al.* [YeSc64] on the emission spectra of Pr^{3+}:LaF_3. Of the many transitions investigated in this work, only the $^3P_0 \to {}^3H_6$ transition will be considered. The $^3P_0$ term is an isolated metastable state; the temperature dependent part of its width is therefore given by

$$\Delta\tilde{\nu}(^3P_0) = (1/2\pi c)H(^3P_0)(T/\Theta)^7 I_6(\Theta/T) \tag{4.14}$$

The value of $H(^3P_0)$ is hard to estimate. Since a transition between the $^3P_0$ term and all the crystal field levels of the $3H_6$ term are being measured in the fluorescence spectrum, linewidth data contain the $\Delta\tilde{\nu}(^3P_0)$ width in addition to that of the $^3H_6$ crystal field levels in which we are interested. Yen *et al.* [YeSc64] measured the transitions $^3P_0 \to a^3H_6$, b^3H_6, and c^3H_6 (where the lowercase letters designate the crystal field levels, starting at the lowest). Of these, $^3P_0 \to a^3H_6$ is the sharpest, and since a^3H_6 can decay only via multiphonon processes to the $^3H_5$ term, which lies 1788 cm^{-1} lower, we can assume that its width at $T = 0$ is given only by the strain. This is also indicated by the fact that the shape of this ($^3P_0 \to a^3H_6$) line is much closer to Gaussian than that of all the other lines seen in the spectrum. Its $\Delta\tilde{\nu} = 0.53\ \text{cm}^{-1}$ width at $T = 0$ K is therefore taken as the strain widths and subtracted from all the other experimental linewidths measured for transitions originating from the $^3P_0$ level. Figure 18 shows the temperature dependence of the linewidth of the three lines mentioned earlier. The figure also shows the equations that have been fitted to the data. In all cases the $0.53\ \text{cm}^{-1}$ strain width has been subtracted from the experimental data. For clarity, the data on the 6000 Å line ($^3P_0 \rightarrow b^3H_6$) will be discussed briefly. The b^3H_6 level can decay to the a^3H_6 level, which is $44\ \text{cm}^{-1}$ lower, and this decay is represented by the first term. The b^3H_6 level can also be depopulated by making transitions to the c^3H_6 level, which is $35\ \text{cm}^{-1}$ above it. Finally, the Raman process must be taken into account. In principle, of course, phonon-induced transitions to the other $(i = \text{d} - m)$ crystal field levels should also be considered, but because of the exponential in the distribution function they contribute very little to the linewidths and are therefore neglected. It can be seen that the equations fit the data quite nicely. It has to be kept in mind that in this procedure the temperature dependence of the $^3P_0$ state is always contained in the measured width. The good agreement between the measured linewidth and the calculated one suggests that the temperature-dependent part of the $^3P_0$ level width contributes little to the actual widths; indeed, estimates show that it can give at most $4\ \text{cm}^{-1}$ at 300 K, which is almost an order of magnitude smaller than the measured widths.

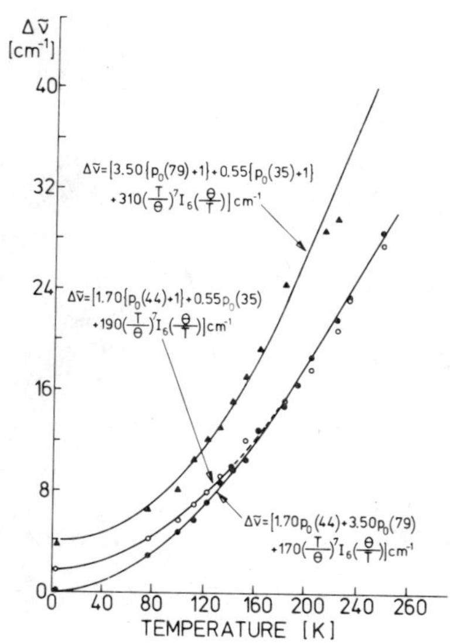

Fig. 18. Temperature dependence of the linewidths $\Delta\tilde{\nu}$ of three strong lines of the $^3P_0 \rightarrow {}^3H_6$ transition in Pr^{3+} : LaF_3. ● indicates 5985 Å; ○, 6000 Å; ▲, 6014 Å; $\Delta\tilde{\nu}^{\text{strain}} = 0.53\ \text{cm}^{-1}$. Solid lines are calculated with the functions shown in the diagram. The experimental points have an accuracy of 10% [YeSc64].

The residual linewidth observed in the optical absorption and emission spectra is due to strains in the crystal and these strains produce an inhomogeneously broadened line. Various techniques employing lasers (optical-free induction decay [GeMa76], photon echos [ChHa76, TaSz74, AbKu66, Ta76], fluorescence line narrowing [Er75], enhanced and saturated absorption spectroscopy [Er77]) have made it possible to study the low temperature homogeneous linewidth. In all these experiments $Pr^{3+}:LaF_3$ crystals were investigated, and thus the data can be compared with the experiments of Yen *et al.* [YeSc64]. The data obtained at temperatures higher than 5 K are consistent with the assumption that the linewidths are caused by electron–phonon interaction. Below 5 K, however, temperature-independent widths of 830 kHz (2.7×10^{-5} cm^{-1}) for the $^1D_2 \rightarrow {}^3H_4$ transition and of 3 MHz (1×10^{-4} cm^{-1}) for the $^3P_0 \rightarrow {}^3H_4$ transition in $Pr^{3+}:LaF_3$ have been found. These widths are larger than expected from radiative and multiphonon decay and have not yet been satisfactorily explained. The most likely explanation seems to be a relaxation via second order hyperfine interactions [Bl73]. This interpretation is supported by the observation of interference effects in the photon echos [ChHa76], which can be explained qualitatively by considering hyperfine interactions.

The width of the $^1D_2 \rightarrow {}^3H_4$ transition was recently also measured by enhanced and saturated absorption spectroscopy [Er77]. Erickson also measured a width of 800 kHz but after deconvolution of the instrumental linewidth arrived at an inherent width for the $^1D_2 \rightarrow {}^3H_4$ transition of (200 ± 50) kHz (6.6×10^{-6} cm^{-1}). This width can be satisfactorily explained by nuclear dipole superhyperfine interaction. It thus seems that now one is able to measure the intrinsic width of optical lines in RE crystal spectra.

Electron–phonon coupling also leads to an energy shift ΔW as a function of temperature; the formula for this energy shift contains matrix elements similar to that for the linewidth, and the result is [YeSc64]

$$\begin{aligned}\Delta W_i(T) = &-\bar{H}_i\left(\frac{T}{\Theta}\right)^4 \int_0^{\Theta/T} dx\,\frac{x^3}{e^x - 1}\\ &+ \sum_{j<i} \bar{K}_{ij}\left(\frac{T}{T_{ij}}\right)^2 P\int_0^{\Theta/T} dx\,\frac{x^3}{e^x - 1}\,\frac{1}{x^2 - (T_{ij}/\Theta)^2}\\ &- \sum_{j>i} \bar{K}_{ij}\left(\frac{T}{T_{ij}}\right)^2 P\int_0^{\Theta/T} dx\,\frac{x^3}{e^x - 1}\,\frac{1}{x^2 - (T_{ij}/\Theta)^2}\end{aligned} \tag{4.15}$$

Here P means the principal value of the integral, $x = W/k_B T$, and $T_{ij} = W_{ij}/k_B$. The first term is, apart from a numerical constant, the expression for the total heat of the crystal under consideration [Ki71]. In eq. (4.15) the first term is due to the Raman process and the next two are caused by direct processes. In a case where the last two terms are negligible, we therefore obtain a linear relation between the total heat of a crystal and the line position, both taken as a function of the temperature [ImYe64].

These last two terms are nearly zero if we are dealing with a situation where one level is so far (meaning several times $k_B\Theta$) removed from the next level that no direct processes can take place. This is the case, e.g., in Eu^{3+} where the $^5D_0$ term lies 1700 cm^{-1} below the $^5D_1$ term and many times 10^3 cm^{-1} above the $^7F_6$ term. The effect of the phonon field on the $^5D_0$ term is then given primarily by the Raman process, which in the energy of the absorption or emission line manifests itself in the first term of eq. (4.15). Figure 19 shows a plot of the line position of the $^5D_0$ term in $Eu^{3+}:YCl_3{\cdot}6H_2O$ and $Eu^{3+}:Y_2O_3$ as a function of the total energy U_G of the undoped compounds, namely $YCl_3{\cdot}6H_2O$ and Y_2O_3. The obvious linear dependence between the two quantities justifies the foregoing approach. This way of checking the interaction between the phonon system and the system of 4f electrons is especially useful because it does not make use of any model for the form of the phonon spectrum.

To emphasize this fact we show in Fig. 20 the data for the temperature dependence of the line shift and a fit to these data using the Raman term of

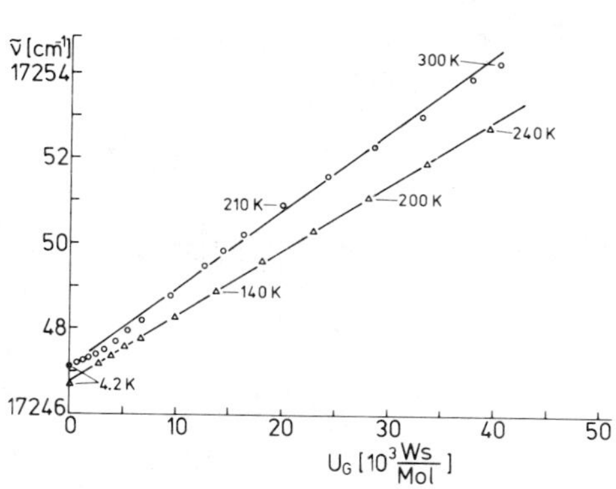

Fig. 19. Line position $\tilde{\nu}$ of the $^7F_0 \rightarrow {}^5D_0$ transition in △, $Eu^{3+}:LuCl_3 \cdot 6H_2O$ (Eu : Lu = 1 : 33), and ○, $Eu^{3+}:Y_2O_3$ (Eu : Y = 1 : 12.5), as a function of the internal energy U_G in $LuCl_3 \cdot 6H_2O$ and Y_2O and Y_2O_3 [HiHü70].
The full curves are straight lines.

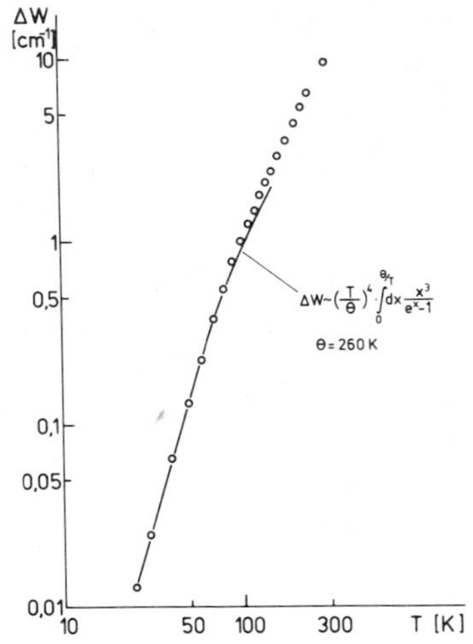

Fig. 20. Line shift ΔW of the $^7F_0 \to {}^5D_0$ transition in $EuCl_3 \cdot 6H_2O$ with respect to the position at 4.2 K; the full curve is a fit to a Raman process with $\Theta = 260$ K [HiHü70] (see eq. (4.15)).

eq. (4.15). The best fit using a Debye spectrum for the phonons is obtained with a Debye temperature of $\Theta = 260$ K, but it can be seen that the curve represents the experimental data only up to a temperature of about 70 K, indicating that the phonon spectrum is not a Debye spectrum. This does not contradict the data in Fig. 19 because in Fig. 20 an explicit model for the phonon distribution had to be assumed.

The foregoing considerations lead to some conclusions about the effects caused by the interaction of the 4f electrons with the acoustical phonons. At low temperature the lowest level of a particular crystal field multiplet will always be relatively long-lived because it can decay only by spontaneous radiative and phonon emission, where the latter is a multiphonon process (assuming that the energy separation to the next crystal field multiplet exceeds the Debye temperature, which is generally the case). The higher levels of a crystal field multiplet can decay, even at $T = 0$ K, to the lower ones by one phonon emission and are therefore much shorter lived. At $T = 0$ K they show a finite width, with the magnitude of these widths increasing with increasing energy separation from the lowest crystal field level of the particular crystal field multiplet. This increase occurs because the higher a crystal field state is, the more possibilities it has to decay to lower ones. The emission of light is mostly observed only from the lowest states of a crystal field multiplet. Especially favorable in this respect are those states which are singlets, like the $^3P_0$ term in Pr^{3+} and the $^5D_0$ term in Eu^{3+}. Most prominent in this respect is the $^4F_{3/2}$

state in Nd^{3+}, which decays by a 1.06 μm photon to the $^4I_{11/2}$ state. This transition is one of the most used laser transitions, especially in the crystal Nd^{3+}:YAG (YAG = yttrium aluminium garnet = $3Y_2O_3 \cdot 5Al_2O_3$), Since this transition has as the final state an excited state ($W(^4I_{11/2}) = 2000\ \text{cm}^{-1}$) that even at room temperature is barely thermally populated, it is very suitable for laser action because an inversion population is easily achievable. In addition, there are a great number of other quite closely spaced Nd terms above the $^4F_{3/2}$ term, which can be easily pumped and which then decay via phonon emission into the $^4F_{3/2}$ term as the final state, thus producing a highly populated $^4F_{3/2}$ state (see Chapter 11 for details).

4.4 Effects of Optical Phonons

The depopulation of an excited state of an RE ion can take place by two processes: by direct transition to a lower state, where the energy is emitted as radiation (light); and by phonon-assisted transitions to a lower state, where the energy is taken up by the crystal in the form of phonons. Since this second class of processes usually involves transition energies larger than 1000 cm^{-1}, high energy phonons (that is, optical phonons) are necessary. The total lifetime of a state is then determined by the probability for radiative and nonradiative processes

$$1/\tau = w^{R} + w^{NR} \tag{4.16}$$

In the Section 4.3 it was shown that transitions within a crystal field multiplet are mediated by acoustical phonons, and that the lifetimes governing these processes are 10^{-10} sec or shorter. By contrast, the measured optical lifetimes are orders of magnitude longer (10^{-6} sec and longer). Therefore, the emission of light is predominantly observed to originate from the lowest energy level of a particular crystal field multiplet because its lifetime is determined by the modes of deexcitation to the other crystal field multiplets. In considering these deexcitation processes of excited states we have to distinguish three cases. (I) Crystals containing no water molecules, where the REs incorporated are greatly diluted (1% concentration or less); these materials show the strongest radiative processes and are therefore important for laser applications. (II) Crystals with no water molecules, in which the RE ions are nevertheless present in concentrated form; energy exchange between neighboring ions can take place in these crystals (see Chapter 5), and the excitation energy can migrate. (III) Crystals containing water molecules; these materials are of no technical importance because they are generally quite volatile. In addition, the water molecules have very strong, high energy optical vibrations, by which the excited states can give energy to the lattice. Therefore, these crystals show much weaker fluorescence than crystals without water.

Here only case (I) will be dealt with in detail because it readily reveals the physics of the processes involved. In this case, the radiative decay rate of an excited state is determined, by the electric and magnetic dipole transition rates, to all states below it; hence, it can be calculated by the Judd–Ofelt theory. The radiationless decay must take place by multi-phonon processes in almost all cases because the maximum phonon energy in crystals is at most 500 cm^{-1}, whereas the energy gaps that have to be bridged are typically several 1000 cm^{-1}. Although this reasoning is simple, it took a long time for these multiphonon processes to be definitely detected.

The most straightforward approach to obtaining the radiationless decay rate is to calculate the radiative probability of a particular transition, subtract it from the measured transition probability, and thus determine the radiationless contribution. To obtain the total radiation probability of a state, we have to calculate A^{ed} and A^{md} to all the states energetically below the one in question; the radiative lifetime is then given by

$$\tau_j = 1\Bigg/\left(\sum_{i<j} A_{ji}^{\mathrm{ed}} + \sum_{i<j} A_{ji}^{\mathrm{md}}\right) \tag{4.17}$$

The nonradiative lifetime is hard to calculate. The theory of multiquantum emission shows, however, that a simple relationship should hold here; if w_{n0} is the rate for the spontaneous emission of n phonons at $T = 0$ K, the temperature dependence of the transition probability for the simultaneous emission of n phonons becomes [MiDe70, Or75]

$$w_{nT}^{\mathrm{NR}} = w_{n0}^{\mathrm{NR}}(1 - \exp(-\hbar\omega/k_{\mathrm{B}}T))^{-n} \tag{4.18}$$

where $\hbar\omega$ is the energy of the phonon under consideration. The constant w_{n0} is, of course, highly dependent on the order n of the process as well. This means that with increasing magnitude of the energy gap between which a transition takes place, the order of the process gets higher and the radiative becomes more important than the nonradiative decay. Therefore, strong emission of light is always observed from terms that are separated from the next lower term by a large energy gap. Looking at Fig. 5, we see that by this reasoning terms with strong light emission should be $^3P_0$ (Pr^{3+}), $^4F_{3/2}$ (Nd^{3+}), $^5D_0$ (Eu^{3+}), $^5D_4$ (Tb^{3+}). These predictions agree with experimental observations.

A quantitative analysis of the radiative and nonradiative decays of RE doped into Y_2O_3 is available and roughly confirms the theory just outlined. By using the intensity parameters determined by fitting the absorption intensities [Kr66], Weber [We68] calculated, using eqs. (3.58) and (3.62), the total rates for radiative decays of a number of terms that have large energy gaps to the next lowest term for Pr^{3+}, Nd^{3+}, Eu^{3+}, Er^{3+}, and Tm^{3+} in Y_2O_3. Table 30 lists some of the results for Eu^{3+} and Er^{3+}, together with measured total lifetimes. For the $^5D_0$ term of Eu^{3+} the measured lifetimes agree very well with the calculated lifetimes, indicating that nonradiative transitions make only a small contribution to the total lifetime. This is not unexpected since the next term, $^7F_6$, is more than 10,000 cm^{-1} below $^5D_0$, and if we assume a maximum optical phonon frequency of about 500 cm^{-1}, a 20-phonon process would be needed to depopulate $^5D_0$ by nonradiative decay. This process is too slow to contribute to the lifetime. For the $^5D_1$ and $^5D_2$ terms the measured lifetimes are much shorter than the calculated ones. This indicates an appreciable nonradiative decay from these terms. For the $^5D_1$ term this can be confirmed by a measurement of the temperature dependence of the lifetime,

TABLE 30

MEASURED AND CALCULATED LIFETIMES OF Eu^{3+} : Y_2O_3 AND Er^{3+} : Y_2O_3 [a]

Excited term	A_{ed} (sec^{-1})	A_{md} (sec^{-1})	Radiative lifetime (μsec)	Observed lifetime (μsec)
Europium				
$^5D_0$	930	109	963	860 ± 20
$^5D_1$	878	105	1020	160 ± 5
$^5D_2$	777	95	1150	≈ 100 (296 K)
Erbium				
$^4I_{13/2}$	73	59.3	7750 ± 2150	8000 ± 500
$^4I_{11/2}$	135	11.9	6810 ± 2470	3900 ± 400
$^4I_{9/2}$	220	1.3	4520 ± 790	$\gtrsim 200$
$^4F_{9/2}$	1537	14.7	644 ± 140	≈ 30
$^2H_{9/2}$	3998	132.4	242 ± 77	≈ 5
$^4G_{11/2}$	65,170	116.3	15 ± 1	$\gtrsim 2$

[a] Temperature is 77 K except where otherwise noted. From [We68].

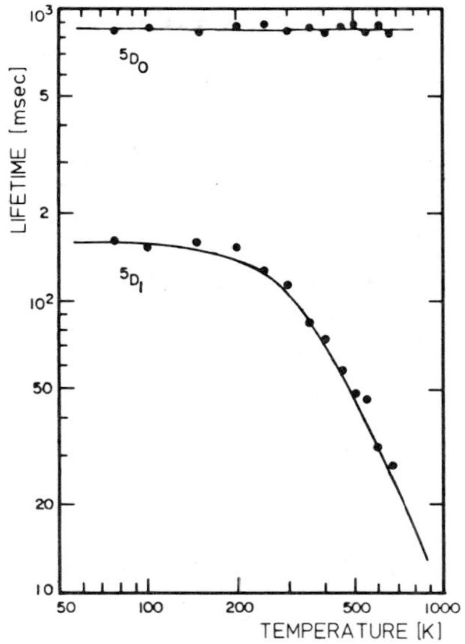

Fig. 21. Temperature dependence of the lifetime of the $^5D_0$, $^5D_1 \rightarrow {}^7F_0$ transitions in $Eu^{3+}:Y_2O_3$.

The full curve for the $^5D_1$ term was calculated with

$$w_{4T}^{NR} = w_{40}^{NR}[1 - \exp(-430/kT)]^{-4}$$

[We68]. The lifetime of the $^5D_0$ term is temperature independent, showing that it is due to spontaneous emission.

which is shown in Fig. 21; in contrast to the lifetime of the $^5D_0$ term, that of $^5D_1$ shows a large decrease (increase of transition probability) with increasing temperature. The curve through the data points in Fig. 21 has been calculated by $1/w_n^{NR}$ from eq. (4.18) using four phonons of 430 cm^{-1}, which are able to bridge the 1715 cm^{-1} energy gap between the $^5D_1$ and the $^5D_0$ terms. To explain the $^5D_1 \rightarrow {}^5D_0$ transition by a multiphonon decay, a higher order process has to be invoked (probably by admixtures of the $^5D_2$ term) because this transition is not allowed in first order.

For $Er^{3+}:Y_2O_3$ the data in Table 30 show a trend similar to that just described for $Eu^{3+}:Y_2O_3$. Usually the calculated radiative lifetime is longer than the measured one, indicating a nonradiative transition contribution. The notable exception is the $^4I_{13/2}$ term, which has a large energy gap and therefore seems to decay only radiatively. Note that as long as there is a term with $\Delta J = 0, \pm 1$ below the decaying one, there is also always a magnetic dipole contribution to the radiative decay probability, because all radiative decay probabilities have to be summed.

The temperature and energy gap dependence of multiphonon transitions has been investigated in a number of hosts by Riseberg and Moos [RiMo67, RiMo68] among others. Their data confirm the previous findings, and by now all the available data show that the highest order of multiphonon processes observable is about six. The rates for spontaneous multiphonon decay seem to depend exponentially on the power of the order of the process because they show a straight line if plotted semi-logarithmically against the energy gap, which the decay has to bridge ($w_{n0}^{NR} \sim e^{-b\Delta W}$, where ΔW is the gap energy). This is evident in Fig. 22, which shows data on Nd^{3+}, Dy^{3+}, Ho^{3+}, and Er^{3+}. Here not the real energy gap, but the gap (ΔW) divided by the maximum optical phonon frequency of the crystal investigated (175 cm^{-1} for $LaBr_3$, 260 cm^{-1} for $LaCl_3$, and 350 cm^{-1} for LaF_3) was used. This way of treating the data eliminates many properties of the phonon system and makes the physical process clearer. It can be seen that the data for $LaCl_3$ and $LaBr_3$, which have the same hexagonal crystal structure, fall on one line, whereas those for LaF_3, which has a different crystal structure, deviate from the former. This shows that in similar crystal fields (as in the case of $LaBr_3$ and $LaCl_3$) the order of the process is the governing factor.

One can rationalize this exponential law as follows [RiMo68, De62, MiDe70]. The electron–phonon coupling that drives the nonradiative decay is proportional to the strain ε (eq. (4.5)). An n phonon process can be obtained by taking the second term in eq. (4.5) to nth order; thus the nonradiative transition probability at $T = 0$ K, namely the prefactor in eq. (4.18), is written as

$$w_{n0}^{NR} = C\varepsilon^n$$

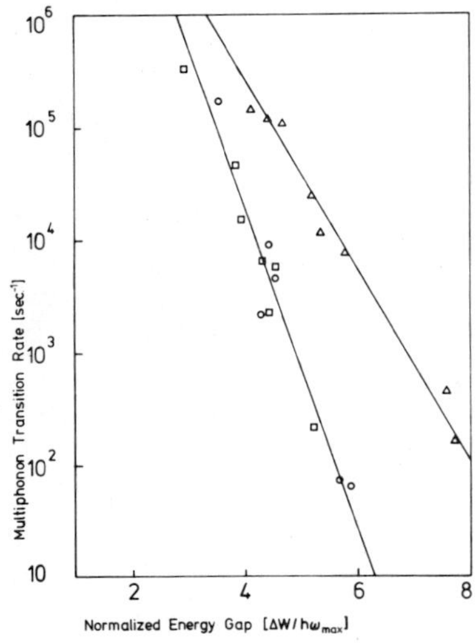

Fig. 22. Zero temperature multiphonon decay rate of RE in lanthanum halogenides (on a logarithmic scale) as a function of a normalized energy gap: $\Delta W/\hbar\omega_{max}$; $\hbar\omega_{max} = 175\ cm^{-1}$ for $LaBr_3$, $\hbar\omega_{max} = 261\ cm^{-1}$ for $LaCl_3$, and $\hbar\omega_{max} = 350\ cm^{-1}$ for LaF_3 [RiMo67].

□ represents $LaCl_3$; ○, $LaBr_3$; △, LaF_3.

Now we assume that the gap energy ΔW can be divided into n phonons $\hbar\omega$ such that $n \cdot \hbar\omega = \Delta W$. This yields

$$w_{n0}^{NR} = C\varepsilon^{\Delta W/\hbar\omega}$$

$$w_{n0}^{NR} = C \exp(-(\Delta W/\hbar\omega)|\ln \varepsilon|)$$

which gives just the behavior seen in the data of Fig. 22.

In an interesting extension of the experiments on the energy gap dependence of the nonradiative transition probability w_{n0}^{NR}, Auzel [Au76] observed the inversion of this process. He chose a typical fluorescent transition, such as the $^4S_{3/2} \rightarrow {}^4I_{15/2}$ transition in Er^{3+}:LaF_3, and measured the intensity of the fluorescent light as a function of the mismatch between the exciting energy and the fluorescent energy. The result is shown in Fig. 23. It is interesting to see that he observed fluorescence from the $^4S_{3/2}$ term, even if he excited with an energy a couple of hundred wave numbers smaller than the term energy, namely in a region where no single ion energy of Er^{3+} can be found. Under the experimental conditions chosen the $^4S_{3/2}$ term can be excited only if several phonons are simultaneously annihilated, because the excitation energy was chosen such that the mismatch in energy ΔW was larger than the maximum phonon energy in LaF_3 (350 cm^{-1}). The excitation probability of the $^4S_{3/2}$ term again follows an exponential law $e^{-b\Delta W}$, where ΔW is the energy mismatch between the $^4S_{3/2}$ term energy and the exciting energy, which can be varied because a dye laser was used for excitation. The same kind of behavior was found if excitation was performed with an energy larger than the $^4S_{3/2}$ term energy. This experiment shows that the simple exponential law holds not only for selected optical phonon frequencies but quite generally for phonons contained in a phonon sideband. An experiment like this can be used to determine the electron–phonon coupling.

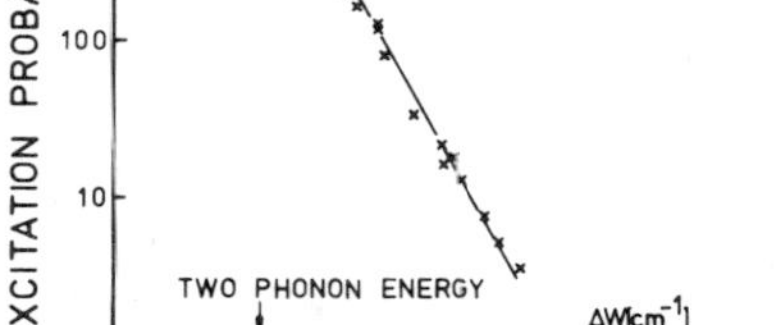

Fig. 23. Excitation probability of the $^4S_{3/2}$ term in Er^{3+} : LaF_3, as monitored by its fluorescence, as a function of the energy mismatch between the exciting radiation and the $^4S_{3/2}$ energy.

The excitation is on the anti-Stokes side (at lower energies) [Au76].

The depopulation of excited states in hydrated crystals has been investigated in detail by Heber [He67]. In such crystals the optical phonons of the water molecules are very effective in depopulating the states by nonradiative multiphonon processes, and since their energies are large ($\sim 2000\,\mathrm{cm}^{-1}$) and their coupling to the RE ions is strong, the nonradiative processes are very much more important in these crystals than in anhydrous materials. Therefore, only a few fluorescent states are known for them (notably $^5D_0$ in Eu^{3+} and $^5D_4$ in Tb^{3+}). In addition, even for these terms, which have very large energy gaps below them, the radiative decay rate is only a few percent of the total decay rate. The data of Heber (which are partly reproduced in Table 31) show an interesting phenomenon. If one forms ratios of the lifetime of the $^5D_0$ term of Eu^{3+} and of the $^5D_4$ term of Tb^{3+} in the hydrated salts, they always give a factor of four. On the other hand, if we compare the lifetimes of different salts for the two REs, we find again that they show very similar ratios, indicating that the lifetimes or transition probabilities can be written in product form:

$$\tau = d b_{\mathrm{RE}} b_{\mathrm{lat}}$$

where d is a constant, b_{RE} a term that depends solely on the RE ion, and b_{lat} a term that depends solely on the lattice. This result is a very nice manifestation of the Born–Oppenheimer postulate, which is always tacitly assumed to hold, that the wavefunctions of a state containing

TABLE 31

LIFETIMES FOR SOME HYDRATED Eu AND Tb SALTS[a] OF THE $^5D_0$ AND $^5D_4$ TERMS, RESPECTIVELY

Compound	Lifetime (μsec)		
	4.2 K	77 K	300 K
$Eu(C_2H_5SO_4)_3 \cdot 9H_2O(^5D_0)$	110	110	111
$Tb(C_2H_5SO_4)_3 \cdot 9H_2O(^5D_4)$	451	429	416
$Eu(C_2H_5SO_4)_3 \cdot 9D_2O(^5D_0)$		>750	>750
$Tb(C_2H_5SO_4)_3 \cdot 9D_2O(^5D_4)$		>1500	>1500
$EuCl_3 \cdot 6H_2O(^5D_0)$	114	114	120
$TbCl_3 \cdot 6H_2O(^5D_4)$	523	493	482
$EuCl_3 \cdot 6D_2O(^5D_0)$		>380	>380
$TbCl_3 \cdot 6D_2O(^5D_4)$		>1100	>1100
$Eu(NO_3)_3 \cdot 6H_2O(^5D_0)$	181	177	164
$Tb(NO_3)_3 \cdot 6H_2O(^5D_4)$	665	652	649

[a] From [He67].

electronic and lattice variables can be written as a product of two functions, one of which depends only on the electron coordinates, and the other only on the lattice coordinates. Large changes in the lifetimes are observed if water is replaced by D_2O. The energy of the vibrations in D_2O is smaller by a factor of $1/\sqrt{2}$ than that in the corresponding hydrated crystals. Therefore, more phonons are needed for the many-phonon nonradiative processes, making the total lifetimes longer.

Finally we note that multiphonon relaxation has also been investigated in RE-doped glasses [LaLo77]. The results are similar to those obtained in crystalline hosts.

CHAPTER 5

Energy Transfer

5.1 Introduction and Theory

So far, the absorption and emission of photon radiation by the RE ions has been considered a one-ion process. This means we have always explicitly assumed that only one ion emits or absorbs a photon. The experimental material available shows, however, that there are a number of cases where two and even more ions can contribute to one absorption or emission process; these experiments will be discussed here. (For a detailed treatment of the material in this chapter see the reviews of Watts [Wa74a, Wa74b], Orbach [Or75], Reisfeld [Re76], and Wright [Wr76] and the literature cited here.) To facilitate our discussion we show in Fig. 24 a number of many-ion processes. The most important class of processes is usually termed energy transfer mechanisms. This means an excited ion, usually called a donor (or a sensitizer), transfers its excitation energy to another ion, called the acceptor (or activator). Less frequent is a second class of processes in which two ions combine to emit or absorb one photon. The following list briefly explains the processes shown in Fig. 24.

(a) In this classical resonant energy transfer process, ion A (the donor) goes from its excited state A* to its ground state A and thereby excites another ion, the acceptor, from its ground state B to its excited state B*.

In this process energy is fully conserved, meaning that the A* → A transition and B → B* transition have the same energy.

(b) In phonon-assisted energy transfer there is a considerable (as compared to the linewidth) mismatch in energy between the acceptor and the donor. Under the conditions shown in the figure, ion B is excited to B* and, to conserve energy, a phonon $\hbar\omega$ is emitted.

(c) This phonon-assisted energy transfer process is analogous to the one in (b); but here the B → B* energy transition is larger than the transition energy A* → A and the missing energy has to be supplied by a phonon annihilation in the system. This process is nonoperative at low temperatures.

(d) Stepwise up conversion is a process in which two successive transitions in the donor system excite one acceptor ion to an energy approximately twice the acceptor energy. A first photon is used to reach the B* state (resonantly or phonon assisted) and a second photon is used to excite the ion from the state B* → C* (again resonantly or phonon-assisted). This means we can produce visible radiation from infrared

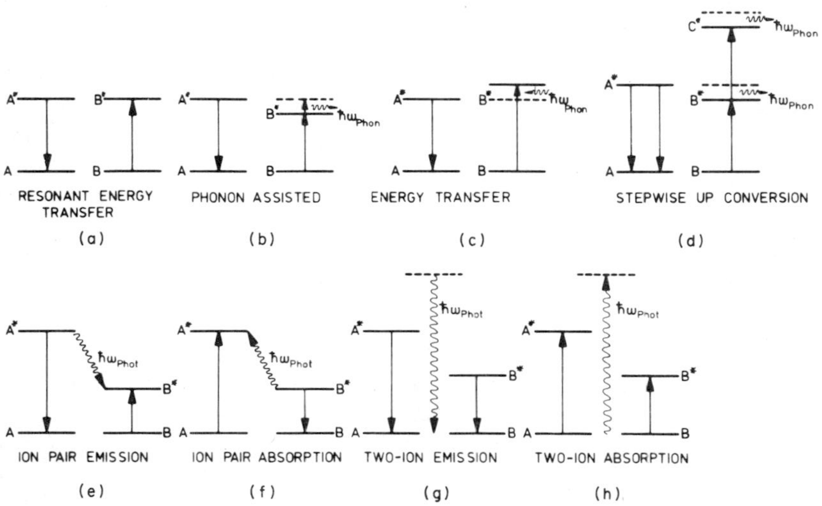

Fig. 24. Diagrams of several energy transfer and two-ion processes.
(a) Resonant energy transfer. (b), (c) Phonon assisted energy transfer with the emission (b) and absorption (c) of a phonon. (d) Stepwise up conversion by phonon assisted energy transfer. (e) Ion pair emission, by which an excited state of ion A* decays and ion B is excited to an excited state B*. (f) Ion pair absorption (reverse of (e)). (g) Two-ion emission: two ions decay simultaneously, emitting a photon with the sum of the energy of their excited states. (h) Two-ion absorption (reverse of (g)).

radiation (emitted, e.g., by the C* → B transition), a useful property in some applications. In passing we also use Figure 24d to explain the quantum counter technique [Bl59]. If the B → B* energy separation is different from that of B* → C*, no radiation can be emitted from C* if the crystal is exposed to radiation of the energy B* → C*. If the B* energy level is populated by shining B → B* radiation onto the crystal, the C* level becomes populated and can emit light. If B → B* is in the infrared region and C* → B in the visible region, this system is called an infrared quantum counter because the low energy B → B* quanta can be made visible by C* → B quanta.

(e) Ion pair emission. The next four processes have in common that two ions act together in order to emit (or absorb) one photon. In this case ion A decays from its excited state A* to its ground state and simultaneously ion B is excited from its ground state to its excited state B*. A photon of energy A* → B* is emitted.

(f) In ion pair absorption the reverse of process (e) happens; absorption takes place from an excited state B* to an excited state A*, which brings ion A from its ground state A to the excited state A* and ion B from the excited state B* to the ground state B.

(g) In two-ion emission two ions, A and B, decay simultaneously from their excited states A* and B*, thereby emitting a photon that has an energy equal to the sum of the excited state energies of the two ions.

(h) In two-ion absorption, the reverse of (g), a photon is absorbed and two ions are thereby brought from the ground state to the excited state.

These processes are not the only ones possible, but they constitute a fair sample of those actually observed. In what follows we shall deal only with processes (a)–(d), because they have been investigated most thoroughly and they have possible importance for the construction of lasers and of infrared detection devices.

The basic theory for energy transfer has been worked out by Förster [Fö65, Fö48] and Dexter [De53, De62].† The rate of transfer of energy from state A to state B is given by

$$w_{\mathrm{AB}} = \frac{2\pi}{\hbar} |\langle \mathrm{A^*B} | \mathscr{H}_{\mathrm{int}} | \mathrm{AB^*} \rangle|^2 \int g_{\mathrm{A}}(W) g_{\mathrm{B}}(W)\, dW \tag{5.1}$$

Here we denote the quantum numbers of the states by A and B, and $\mathscr{H}_{\mathrm{int}}$ is the interaction Hamiltonian for the transition between the two ions. $g_{\mathrm{A}}(W)$ and $g_{\mathrm{B}}(W)$ are the two line shape functions of the transitions

† The problems in the application of that theory to RE systems has been discussed by Heber [HeMu77].

$A^* \rightarrow A$ and $B \rightarrow B^*$. We see that this formula is valid only for resonant energy transfer because the integral gets small very rapidly as a mismatch between the donor and acceptor energies occurs. The transition matrix elements between the ions can be of three types: electric multipole interaction, exchange interaction, and virtual phonon exchange interaction. The last interaction is complicated and can be thought of as coming about by the emission of a phonon by ion A and absorption of this phonon by ion B, whereby an interaction is mediated. So far experiments on REs have been interpreted mostly in terms of electric multipole interactions between the ions (exchange interactions are important for the 3d ions). This means the interaction Hamiltonian can be written and worked out in the same way as for the crystal field interaction in Chapter 3. When discussing experiments, only the range dependence of the interaction, given by

Dipolar–dipolar	$1/(R_{AB})^6$
Dipolar–quadrupolar	$1/(R_{AB})^8$
Quadrupolar–quadrupolar	$1/(R_{AB})^{10}$
Magnetic dipole	$1/(R_{AB})^6$
Exchange	$\exp[-2z\lvert\ln S\rvert]$

(z is the number of ligands and S the overlap integral) is important.

The theory of phonon-assisted energy transfer was first developed by Orbach [Or67] and then worked out in detail by Miyakawa and Dexter [MiDe70] and Orbach [Or75]. The situation is not completely settled since there is still some disagreement about how to calculate the interaction matrix element. We give only the formula which reads [Or75]

$$w_{AB} = \frac{2\pi}{\hbar} |\langle A^*B|\mathscr{H}_{int}|AB^*\rangle|^2 K \begin{bmatrix} p(\omega) \\ p(\omega) + 1 \end{bmatrix} \int g_A(W) g_B(W \overset{+}{-} \hbar\omega)\, dW \tag{5.2}$$

where K is now a combination of electron–phonon matrix elements (the exact evaluation of which differs in [MiDe70] and [Or75]) and $p(\omega)$ is the Bose function for the phonon occupation; $p(\omega)$ and $(W + \hbar\omega)$ stand for process (c), and $p(\omega) + 1$ and $(W - \hbar\omega)$ for process (b).

Although the exact nature of this equation is still being debated, there is no argument about the dependence of the energy transfer rate on the energy mismatch (called the energy gap ΔW in Section 4.4), if more than one phonon is needed, and on its temperature dependence. As Miyakawa and Dexter [MiDe70, Or75] have shown, we obtain the same dependence as in the case of the nonradiative decay rates of excited crystal field states. Thus we can write the energy transfer rate as a function of the energy mismatch ΔW if the mismatch exceeds the Debye temperature:

$$w_{\mathrm{AB}} = \mathrm{B} \exp(-b\Delta W) \tag{5.3}$$

and as a function of temperature

$$w_{\mathrm{AB}}(T) = w_{\mathrm{AB}}(T = 0)(1 - \exp(-\hbar\omega/k_{\mathrm{B}} T))^{-n} \tag{5.4}$$

where $\hbar\omega$ is the average phonon energy needed to bridge the energy gap in an n-phonon process ($\Delta W = n\hbar\omega$).

The range dependence of the transfer rate for nonradiative transfer is given mainly by the interaction matrix element and is therefore the same as in resonant energy transfer. There are, however, a few weakly distant dependent terms in the electron–phonon matrix elements.

The question now arises how to obtain w_{AB} from an experiment. For the donor ions the possibility to transfer energy to the acceptor ions constitutes an additional possiblity to decay from an excited state (in addition to radiative and nonradiative decay). This modifies the donor excited state lifetime. A good way to investigate energy transfer is therefore to measure the donor fluorescence lifetime. This means, we need equations that relate w_{AB} to, e.g., the time dependence of that fluorescence lifetime. To obtain them we assume, for a crystal of volume V, that there is a donor concentration c_{A} and an acceptor concentration c_{B}; that the intrinsic donor lifetime is τ_{A} (measured for small donor concentrations and in the absence of any acceptor); and that the donor–acceptor transfer rate has the form $w_{\mathrm{AB}} = a_s/R^s$. Very often a so-called critical transfer radius is introduced in the form $1/\tau_{\mathrm{A}} = a_s/R_0^s$, which means that it is the distance at which $w_{\mathrm{AB}} = 1/\tau_{\mathrm{A}}$. The donor–acceptor transfer rate can then be expressed in the form $w_{\mathrm{AB}} = (R_0/R)^s \cdot 1/\tau_{\mathrm{A}}$. The critical transfer radius R_0 can be related to a critical concentration by $c_0^{-1} = 4\pi R_0^3/3$. Finally, we have to take into account that a donor can decay not only radiatively and nonradiatively, or by transferring energy to the acceptor system, but also by exchanging energy with another donor. The last process, which depends on the donor concentration, can be thought of as energy migration or diffusion in the donor system and is therefore described by a diffusion constant D.

Now the rate equation for the population in the donor system, taking into account the various decay channels, is set up; from it, the time dependence of the donor fluorescence intensity can be calculated.

In cases where interaction between donors can be neglected and we have to take account only of the donor–acceptor transfer and the lifetime of the donors, the time dependence of the donor fluorescence signal intensity is given by

$$\frac{I_1(t)}{I_1(0)} = \exp[((-t/\tau_A) - (c_B/c_0)\Gamma(1 - (3/s))(t/\tau_A)^{3/s}] \tag{5.5}$$

Here s is the exponent of the range interaction (namely, 6 for dipole–dipole, 8 for dipole–quadrupole, and 10 for quadrupole–quadrupole interaction), and Γ is the Γ function. From a fit of this equation to the actual observed decay signal, in principle the value for s can be determined and hence the nature of the interaction. We note that the curves do not depend strongly on s and that such determinations therefore contain some ambiguity. However, most experiments have been analyzed by using $s = 6$, assuming that the dipole–dipole interaction is the dominant term producing the energy transfer in RE compounds. The decay curves also yield c_0 and τ_A, which enables us to determine w_{AB}.

In the case of larger donor concentrations, exchange of energy between donors becomes operative. This process is by definition a resonant one and therefore quite rapid. The decay curve of the donor fluorescence in this case has been calculated approximately only for dipole–dipole interaction a_6/R^6, yielding

$$\frac{I_2(t)}{I_2(0)} = \exp\left\{(-t/\tau_A) - (c_B/c_0)\Gamma(\tfrac{1}{2})(t/\tau_A)^{1/2}\left[\frac{1 + 10.87y + 15.5y^2}{1 + 8.743y}\right]^{3/4}\right\} \tag{5.6}$$

with $y = D\tau_A R_0^{-2}(t/\tau_A)^{2/3}$. For small t this formula approaches that for zero diffusion and is therefore nonexponential.

For large t eq. (5.6) leads to an exponential decay, where the decay constant is given by

$$\frac{1}{\tau} = \frac{1}{\tau_A} + 4\pi D c_B R_s \tag{5.7}$$

and $R_s = 0.68(a_6/D)^{1/4}$. (This formula does not follow exactly from eq. (5.6) because it was obtained by an approximate solution of the rate equation. Equation (5.7) is rather an exact solution for large t [deGe58].)

With respect to the donor concentration c_A we can distinguish three cases. For small donor concentration the donor–donor interaction is small and the donor decay time dependence is given by eq. (5.5). This curve begins with a nonexponential part and for large times the decay is exponential with τ_A. With intermediate concentrations c_A, migration (diffusion) becomes important. Again at small t there is a nonexponential decay, and for large t we find an exponential diffusion-limited decay. For very large donor concentrations almost every acceptor may be the nearest neighbor to a donor, and we have an exponential decay where the decay constant is given by the nearest neighbor donor–acceptor transfer rate. A few examples of energy transfer studies will be given in the next section.

Cooperative emission and absorption processes will not be discussed explicity; the reader is referred to the original literature. The theory for these processes has been worked out by Dexter [De62, AlDe73] and processes of type (e) have been observed by Feofilov and Trofimov [FeTr69] and van der Ziel and Van Uitert [ZiUi69, ZiUi73], who have also observed process (f), as well as Bilak *et al.* [BiZv71]; process (g) has been seen by Nakazawa and Shionoya [NaShi70] and process (h) by Varsanyi and Dieke [VaDi61].

5.2 Energy Transfer in Rare Earth-Doped YF_3 and Y_2O_3

There exist many investigations of energy transfer in RE solids that initiated in an attempt to obtain more powerful lasers. We shall use two examples to show how the formula in Section 5.1 can be used to analyze data and to show what the order of magnitude values for the various parameters are. Crystals whose optical properties suggest that they may be suited for laser applications and that were used as examples in previous sections are dealt with.

The change from the behavior with no donor–donor interaction to that with strong donor–donor interaction can be seen, e.g., in the system of YF_3 doped with Yb and Ho [WaRi72] (see, e.g., [OsZi71] for other acceptors). Here Yb acts as a donor (via the $^2F_{5/2} \rightarrow {}^2F_{7/2}$ transition) and Ho is the acceptor. The acceptor can take up energy by a two-step process where the first is $^5I_8 \rightarrow {}^5I_6$, a phonon-assisted transition, and the second is $^5I_6 \rightarrow {}^5S_2$, a resonant energy transfer process. Since the first step is phonon assisted, we can expect a relatively small donor–acceptor transfer rate. An example of donor decay curves for small donor concentrations is shown in Fig. 25. We see that for small acceptor concentrations the curves are exponential over the whole time scale, yielding the donor decay time τ_A. For larger acceptor concentrations c_B, we note the anticipated nonexponential decay at the beginning of the decay curve and the exponential

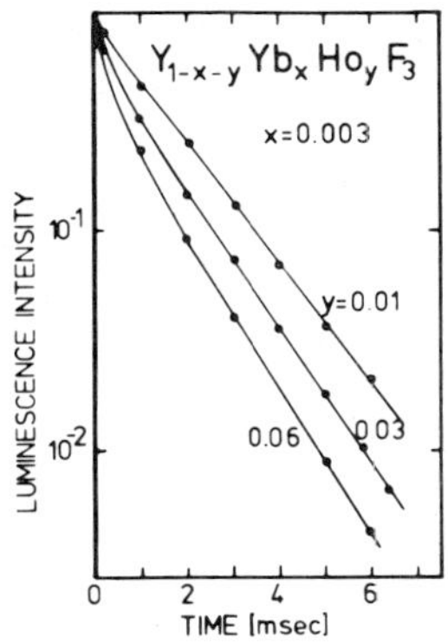

Fig. 25

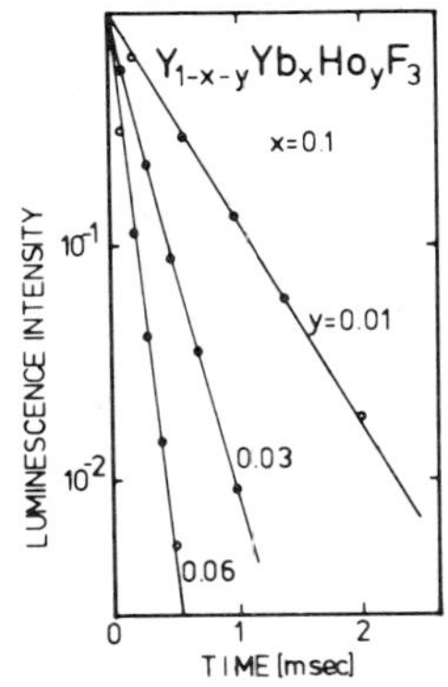

Fig. 26

Fig. 25. Fluorescence decay of Yb^{3+} in $Y_{1-x-y}Yb_xHo_yF_3$.
The parameters of the curves are $x = 0.003$ and, from top to bottom, $y = 0.01$, $y = 0.03$, and $y = 0.06$, respectively. The points were fitted with eq. (5.5) and $s = 6$ [WaRi72].

Fig. 26. Fluorescence decay of Yb^{3+} in $Y_{1-x-y}Yb_xHo_yF_3$.
The parameters of the curves are $x = 0.1$ and, from top to bottom, $y = 0.01$, $y = 0.03$, and $y = 0.06$, respectively. The points were fitted with eq. (5.6) [WaRi72].

decay for large times, again varying with approximately the donor decay time τ_A. These curves were fitted with eq. (5.5) by using the exponent for the dipole–dipole interaction ($s = 6$). The good agreement of the measured decay curves with the calculated points indicates that the dipole interaction is probably a governing process. Figure 26 shows an example with large donor concentrations. The curves which are distinctly different from those in Fig. 25, were fitted with eq. (5.6). Now the exponential tail of the curves is a function of the acceptor concentration c_B, which it should be according to eq. (5.7). Analysis of the curves for a number of donor/acceptor combinations yields $\tau_A = 1.7 \times 10^{-3}$ sec, $R_0 = 6$ Å, and $a_6 = 1.8 \times 10^{-41}$ cm^6/sec. a_6 is approximately independent of donor and acceptor concentrations. It has a relatively small value (due to the nonresonant nature of the energy transfer). Krasutsky and Moos [KrMo73] find, for energy transfer in $LaCl_3$ between Pr and Nd ions in the infrared spectral region, a coupling constant $a_6 = 6 \times 10^{-38}$ cm^6/sec, considerably larger than in the present case showing the large spread possible for the strength of the coupling. The numbers for YF_3 enable us to calculate $w_{AB} \sim 10^4$ sec^{-1}—again a relatively small number. In Table 32 we list the average donor distance (c_A^{-3}) calculated from the donor concentration and compare it with the diffusion length of the excitation

TABLE 32

ENERGY TRANSFER PARAMETERS FOR $Y_{1-x-y}Yb_xHo_yF_3$ ($y = 0.01$)[a]

x	D (cm^2/sec)	$(D\tau_A)^{1/2}$ (Å)	$c_A^{-1/3}$ (Å)
0.1	1.6×10^{-11}	18	8
0.03	1.1×10^{-11}	17	12
0.01	2.9×10^{-12}	6	17
0.003	2.7×10^{-14}	0.8	25

[a] From [WaRi72].

energy namely the distance the excitation energy travels within one donor lifetime $(D\tau_A)^{1/2}$. We see that at the smallest donor concentrations the diffusion length is much shorter than the average donor–donor distance, indicating that the energy diffusion is negligible. Between $c_A = 0.03$ and $c_A = 0.1$ the diffusion constant changes very little, indicating that at these concentrations the energy transfer is dominated by the nearest neighbor donor–acceptor transfer time.

Extensive data also exist on nonradiative energy transfer of REs in Y_2O_3 [YaShi72]. In this case the authors assumed that dipole–quadrupole coupling is the dominant process, although in looking at their data we could just as well assume a dipole–dipole or quadrupole–quadrupole coupling. Table 33 presents results for the oxide system. All data were obtained with a donor concentration of about 1% and an acceptor concentration of 4%, which corresponds to an average acceptor–donor separation of 10 Å. The table lists the donors and acceptors, together with the donor and acceptor transitions involved. The energy gap between the donor energy emitted and the acceptor energy is also shown. The energy gap shows a large variation and can therefore be used to investigate the energy gap dependence of the energy transfer rate. The donor decay time τ_A was obtained from eq. (5.5) by setting

$$I_1(\tau_A) = (1/e)I_1(0)$$

An interesting investigation on energy transfer in Y_2O_3 was also performed by Heber *et al.* [HeHe70] who studied the transfer of energy from the C_2 to the C_{3i} sites in $Eu^{3+}:Y_2O_3$.

If the energy transfer rate is plotted as a function of the energy gap, approximately the same exponential behavior is noted as was found

TABLE 33

Observed Donor Fluorescence Decay Times in the Absence (τ_{A_0}) and Presence (τ_A) of Acceptor, and Calculated Phonon-Assisted Energy Transfer Rates for Various Donor/Acceptor Combinations of Rare Earth Ions in Y_2O_3 at 77 K; $c_A = 1\%$, $c_B = 4\%$[a]

Donor ion	Acceptor ion	Transition: Donor	Transition: Acceptor	Energy gap (cm^{-1})	Decay time (μsec): τ_0	Decay time (μsec): τ_e	Energy transfer rate ($R = 10$ Å) (sec^{-1})
Sm	Eu	$^4G_{5/2} \rightarrow {}^6H_{5/2}$	$^7F_0 \rightarrow {}^5D_0$	600	290 ± 10	87 ± 7	1.7×10^3
Eu	Yb	$^5D_0 \rightarrow {}^7F_6$	$^2F_{7/2} \rightarrow {}^2F_{5/2}$	1670	880 ± 20	650 ± 20	1.7×10
Eu	Yb	$^5D_1 \rightarrow {}^5D_0$	$^2F_{7/2} \rightarrow {}^2F_{5/2}$	1225	77 ± 5	69 ± 5	4.0×10
Eu	Yb	$^5D_2 \rightarrow {}^5D_1$	$^2F_{7/2} \rightarrow {}^2F_{5/2}$	1935	29 ± 2	20 ± 2	1.1×10^3
Tb	Yb	$^5D_4 \rightarrow {}^7F_0$	$^2F_{7/2} \rightarrow {}^2F_{5/2}$	4200	990 ± 20	850 ± 20	2.3
Ho	Sm	$^5S_2 \rightarrow {}^5I_4$	$^6H_{5/2} \rightarrow {}^6H_{13/2}$	190	91 ± 2	5.5 ± 0.5	5.9×10^4
Ho	Tm	$^5S_2 \rightarrow {}^5I_7$	$^3H_6 \rightarrow {}^3F_4$	480	91 ± 2	13.6 ± 0.5	1.8×10^4
Ho	Yb	$^5I_4 \rightarrow {}^5I_8$	$^2F_{7/2} \rightarrow {}^2F_{5/2}$	2610	47 ± 5	32 ± 5	4.6×10^2
Er	Yb	$^4S_{3/2} \rightarrow {}^4I_{13/2}$	$^2F_{7/2} \rightarrow {}^2F_{5/2}$	1070	88 ± 5	33 ± 2	8.4×10^3
Tm	Yb	$^1G_4 \rightarrow {}^3H_5$	$^2F_{7/2} \rightarrow {}^2F_{5/2}$	1840	42 ± 5	30 ± 2	5.3×10^2

[a] From [YaShi72].

previously (see Fig. 22) for the decay constant of excited levels, showing that the same sort of process is operative. In addition, the temperature dependence of the energy transfer rate has been investigated in a few cases and here too we find the same dependence as was found for nonradiative decay, namely, a process activated by a number of phonons. These data confirmed that nonresonant or phonon-assisted energy transfer follows the same pattern as the nonradiative decay of excited levels.

In order to assess the range of validity of the Förster–Dexter theory [Fö48, De53] especially in the case of small acceptor concentrations, we have to look closer into the energy transfer process [HeMu77]. An excited donor transfers its energy in the donor system until a donor is reached which has in its sphere of influence an acceptor to which it transfers its excitation energy. Usually it is assumed that the donor–donor transfer is fast (because it is a resonant process) as compared to the donor–acceptor transfer. It was however shown in Section 4.3 that the intrinsic linewidth of the RE levels at low temperatures is only of the order of 10^{-6} cm^{-1} [Er77], whereas the inhomogeneous width is of the order of 1 cm^{-1} [YeSc64]. Thus the overlap of the linewidth of different donor ions is negligible, and the donor–donor transfer rate gets slow (the excitation energy is highly localized) and can be the determining factor in the donor–acceptor energy transfer [HeMu77]. The donor–donor energy transfer will under these circumstances also be phonon assisted as shown experimentally by Flach *et al.* [FlHa75] and theoretically by Holstein *et al.* [HoLy76].

The localization of the excitation energy in systems with an inhomogeneously broadened line if the interaction between ions falls off more rapidly than r^{-3} has been predicted by Anderson [An58] ("Anderson localization"). This phenomenon has been observed in $Sm^{3+}:CaWO_4$ [HsPo75].

CHAPTER 6

Magnetic Interactions

6.1 Introduction

Thus far we have neglected magnetic interactions between the RE ions. However, any ensemble of paramagnetic ions will attain a magnetically ordered state if their temperature is lowered sufficiently. The question then arises how this magnetically ordered state changes the crystal field energy levels, and what information about this state can be obtained from an investigation of the energy levels by optical spectroscopy. We emphasize that these effects play only a minor role for RE compounds because their magnetic interactions are relatively small. Rare earth magnetic dipole interactions are of the order of a few wavenumbers or less and therefore lead to ordering temperatures in the degree range. Exchange interactions between RE ions are relatively small also, because of the localized nature of the 4f orbitals; hence these interactions cannot raise the ordering temperature much. An interesting class of materials, however, are those which are a mixture of oxides of RE ions and 3d ions, like the RE iron garnets, RE orthoferrites, or RE orthochromites. The delocalized nature of the 3d orbitals in these substances leads to large exchange interactions on the RE ions that have been investigated in detail by optical means. However, it must be admitted that not much significant information about magnetism has been obtained in this way, although some interesting phenomena have been observed. For a review of this field see the articles by Leask [Le68] and Hüfner [Hü71, Hü72, Hü76].

6.2 General Considerations

Most of the commonly known magnetic materials are those containing 3d ions: iron metal, ferrites, chromites, etc. Therefore, theories treating magnetic phenomena are tailored to these materials rather than RE ions. Unlike the RE-containing compounds, those with 3d ions have as their main characteristic the so-called quenching of the orbital angular momentum. This means that the magnetic interactions are dominated by spin interactions. In contrast, for RE ions, with their large spin–orbit interaction, the states are characterized by the total angular momentum, which contains large orbital contributions. In these circumstances, the interaction is much more complicated, being very anisotropic and containing many more interaction constants. We shall start by presenting the Hamiltonian for the 3d ions and shall then introduce that for the RE ions.

If the magnetic ions in a compound have only spin angular momentum, the interaction Hamiltonian between two such ions has the form [Ba71, Wo71]

$$\mathscr{H}_{12} = J\mathbf{S}_1 \cdot \mathbf{S}_2 + \mathbf{d} \cdot (\mathbf{S}_1 \times \mathbf{S}_2) + \mathbf{S}_1 \cdot D_{\mathrm{E}} \cdot \mathbf{S}_2 + j(\mathbf{S}_1 \cdot \mathbf{S}_2)^2 + \mathbf{S}_1 \cdot D_{\mathrm{d}} \cdot \mathbf{S}_2 \tag{6.1}$$

Here J is the Heisenberg exchange integral, $\mathbf{d}$ the antisymmetric exchange constant, D_{E} the anisotropic bilinear exchange constant, j the constant of the quadratic exchange, and D_{d} the measure of the classical magnetic dipole interaction. We follow Baker [Ba71] and Wolf [Wo71] in the sign convention for the first term, although it is not the usual one (in addition, a factor of two is missing as compared to the more common Hamiltonian). In the convention of eq. (6.1) an antiferromagnetic exchange interaction is described by a positive J and a ferromagnetic one by a negative J. The last term is the only one that can be calculated immediately from the magnetic moments $\boldsymbol{\mu}_1$ and $\boldsymbol{\mu}_2$ of the two ions if their mutual separation $\mathbf{r}$ is known:

$$\mathbf{S}_1 \cdot D_{\mathrm{d}} \cdot \mathbf{S}_2 = \frac{\boldsymbol{\mu}_1 \cdot \boldsymbol{\mu}_2}{r^3} - \frac{3(\boldsymbol{\mu}_1 \cdot \mathbf{r})(\boldsymbol{\mu}_2 \cdot \mathbf{r})}{r^5} \tag{6.2}$$

All the other contributions can at best be estimated from models.

In a crystal a particular ion interacts with all the other ions carrying a magnetic moment via the Hamiltonian (6.1). To solve this complicated many-body problem in the simplest way, the molecular field model is used. Considering for a moment only the first term, this means

$$\mathscr{H}_j = \mathbf{S}_j \cdot \sum_i J_{ij} \mathbf{S}_i \tag{6.3}$$

and replacing $\mathbf{S}_i$ by its thermal expectation value $\langle \mathbf{S}_i \rangle$, we obtain

$$\mathscr{H}_j = \mathbf{S}_j \cdot \langle \mathbf{S}_i \rangle \sum_i J_{ij} \tag{6.4}$$

To make this equation more readable it is useful to write it in the form of a magnetic moment times a magnetic field and to call this field the exchange (molecular) field:

$$\mathscr{H}_j = \boldsymbol{\mu}_{sj} \cdot \mathbf{H}^{\mathrm{ex}} \tag{6.5}$$

Since $\boldsymbol{\mu}_{sj} = g\mu_{\mathrm{B}}\mathbf{S}_j$ we obtain

$$g\mu_{\mathrm{B}}\mathbf{H}^{\mathrm{ex}} = \langle \mathbf{S}_i \rangle \sum_i J_{ij} \tag{6.6}$$

as the definition of the exchange field H^{ex}.

Many experimental observations of magnetically ordered RE ions can be interpreted within the molecular field model. This means that effects are observed that are analogous to those seen in the ordinary Zeeman measurements. Since the states of the RE ions are characterized by the total angular momentum and not by the spin, it is useful to write the molecular field Hamiltonian in terms of the total angular momentum

$$\mathscr{H} = g_J \mu_{\mathrm{B}} \mathbf{J} \cdot \left(\frac{2(g_J - 1)}{g_J} \mathbf{H}^{\mathrm{ex}} \right) \tag{6.7}$$

where g_J is the Landé g_J-factor of the RE state under consideration.

Although usually not mentioned in this context, the time domain is important for determining whether an optical experiment can be interpreted within the molecular field theory or not. Ordinarily it is assumed that thermal averaging of the magnetic moments is fast as compared to the measuring time of a specific experiment, which manifests itself, e.g., in the linewidth. As a consequence, in a paramagnetic crystal in zero external field we see no effect of the magnetic moments on the optical levels, because macroscopically speaking (as verified, e.g., by a magnetization measurement) there is no magnetic moment. The lifetimes of the optical states can, however, be so short that the optical experiment measures not the thermal average of the ensemble of the magnetic moments, but rather the instantaneous configuration of the moments surrounding a particular ion. Then the optical experiment permits us to deduce the interactions between neighbors, as we shall see. We can also roughly predict when this will be the case. If we assume that much of the thermal averaging of the moments is done by their off-diagonal components, we may expect that the averaging will not occur in cases where these components are small. Thus we expect the molecular field approach to break down in substances where the interactions are Ising-like, namely of the $J_{1z} \cdot J_{2z}$ type; this is indeed observed.

The Hamiltonian (6.1) is rigorously valid for ions that have states in the crystal with quenched orbital angular momentum. For simplicity it is often also used for RE ions. A correct treatment for these ions has to consider the effects produced by the anisotropy of their charge distribution. This leads to the development of a magnetic interaction between ions that looks very similar to that for the crystal field interaction. There are, however, far fewer results available, and therefore this approach has not had much practical application yet and is not very likely to do so in the future. For this reason we quote only the general equations without going into detail. (But see [vVl62, Le64, Le66a, Ba71, Wo71].) Formally we can derive the exchange interaction between two ions with orbital momentum by developing the potential into spherical harmonics and then introducing tensor operators, to obtain the two particle Hamiltonian

$$\mathscr{H}_{12} = \sum_{kk'qq'} \bar{J}_{kk'qq'} C_{kq}(1) C_{k'q'}(2) \tag{6.8a}$$

Since the states of the RE ions are characterized by the total angular momentum J, it is sometimes convenient to rewrite this equation in spin Hamiltonian notation, using the Racah operator equivalents

$$\mathscr{H}_{12} = \sum_{kk'qq'} J_{kk'qq'} \tilde{O}_{kq}(\mathbf{J}_1) \tilde{O}_{k'q'}(\mathbf{J}_2) \tag{6.8b}$$

which, if only the first term is retained, yields

$$\mathscr{H}_{12} = (g_J - 1)^2 J_{12} \mathbf{J}_1 \cdot \mathbf{J}_2 \tag{6.8c}$$

A case of special interest is that in which one of the two interacting ions is an S state ion. This happens for the RE iron garnets, a class of much investigated mixed RE oxides and iron oxide where the RE ions experience a large exchange interaction coming from the iron ions.

6.3 Experimental Results: Introduction

As mentioned earlier in all RE compounds except RE iron garnets the magnetic interactions experienced by the RE ions are smaller than the crystal field interaction. This makes the description in terms of a spin Hamiltonian formalism possible. This approximation also proved useful for the few cases where the RE garnets have been investigated. We shall now discuss separately experiments where the molecular field approach is useful, where effects due to neighbor configurations have been observed, and finally where the spin-wave theory has been applied. A word of warning: Magnetic phenomena are quite complicated and to treat them thoroughly would require a separate volume. Therefore in what follows we present examples that show most clearly the effects magnetic interactions can have on the optical lines of RE compounds, with no intention of covering the field of magnetism in REs (see, e.g., [El72]).

6.4 Molecular Field Approach: $ErCrO_3$, YbIG

This section comprises experiments that can be described by approximating the magnetic interactions by a magnetic field. The anisotropy of the interaction will not be neglected, but effects of individual neighbors and of the dispersion will. This means we shall assume that the magnetic interactions act like an external magnetic field, which can have different magnitudes in different crystal directions.

One especially instructive experiment was unfortunately performed on the somewhat complicated crystal $ErCrO_3$; nevertheless we describe the experiment here to show the possibilities in this area [CoHü75, Co75a, HaHo75].

$ErCrO_3$ has the perovskite structure. The Cr ions (in which we are not interested) order antiferromagnetically at $T_N = 133$ K and below this temperature produce a magnetic field at the site of the Er ions. The Cr ions are ordered approximately along the $\pm x$ axis, with a small canting of the moments in the x–z plane toward the positive z axis (Γ_4 phase) [Be63]. Magnetic configurations have been classified by Bertaut [Be63] in a way similar to the classification of the crystal structures. This classification of the magnetic structures will not be elaborated on here because we use it only at this point. Figure 27 shows the relative directions of Cr^{3+} and RE^{3+} moments. At a second magnetic ordering temperature T_R, the Cr ions undergo a spin reorientation, after which they point in the $\pm y$

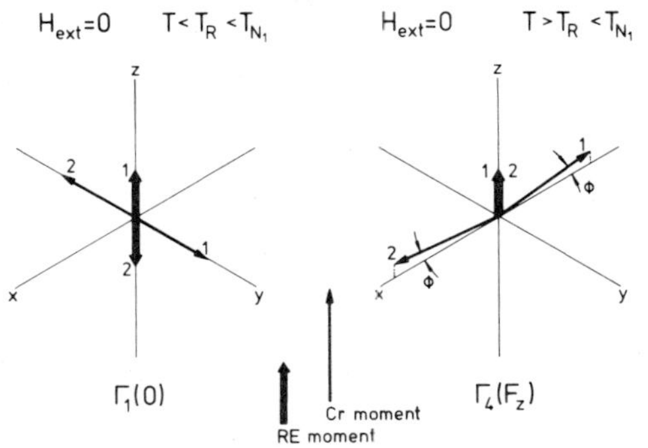

Fig. 27. Relative directions of the magnetic moments of the RE^{3+} and Cr^{3+} ions in $ErCrO_3$.
$\Gamma_1(0)$ is the configuration below the reorientation temperature $\Gamma_R = 9.5$ K, $\Gamma_4(F_z)$ the configuration above it, up to the chromium sublattice ordering temperature of $T_N = 135$ K [CoHü75].

direction without any canting (Γ_1 phase). Below this reorientation temperature the magnetic field experienced by the Er ions is larger than in the Γ_4 phase. This shows that the reorientation of the Cr sublattice is brought about by a competition between the anisotropy energy (the energy it takes to change the Cr moments from the $\pm x$ direction to the $\pm y$ direction) of the Cr sublattices and the thermal energy in the Er system: If the gain in energy resulting from the enlargement of the Er ground state splitting is larger than the loss in anisotropy energy in going from the Γ_4 to the Γ_1 phase, the reorientation phase transition takes place [Wh69]. The effects of the magnetic order on the absorption spectrum in the ${}^4I_{15/2} \rightarrow {}^4S_{3/2}$ transition will be discussed; the relevant data are shown in Fig. 28.

The lower part of the figure shows an energy level diagram for two doublets split by a magnetic field and the allowed optical transition for point symmetry C_{1h}. There is a difference for the optical transitions polarized parallel and perpendicular to the crystallographic z axis. The top part of Fig. 28 shows a spectrum at 4.2 K of the ${}^4I_{15/2} \rightarrow {}^4S_{3/2}$ transition for the two polarizations. The spectra have been taken with zero applied magnetic field. If only the crystal field were acting on the Er ions, the transitions would consist in both polarizations of only two lines (e.g., the $Er(C_2H_5SO_4)_3 \cdot 9H_2O$ spectrum shown in Fig. 8), reflecting the crystal field splitting of the ${}^4S_{3/2}$ excited term (the first excited crystal field level of the ${}^4I_{15/2}$ ground term is far removed from the lowest crystal field level, and transitions from it are therefore not visible at 4.2 K). In

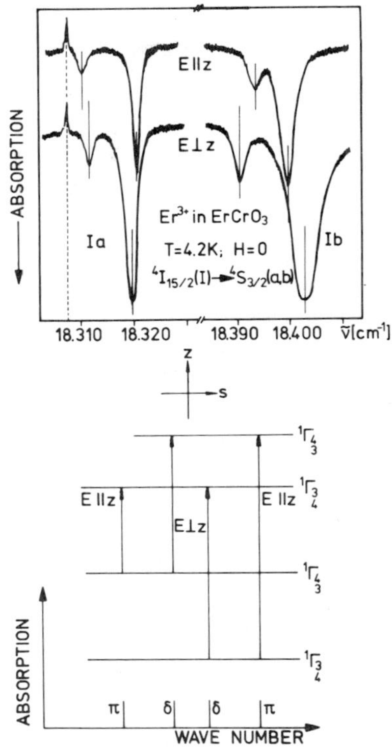

Fig. 28. $^4I_{15/2} \rightarrow {}^4S_{3/2}$ transition in the absorption spectrum of $ErCrO_3$ at 4.2 K for **E** ∥ z and **E** ⊥ z.

Upper part: Absorption spectrum for the two polarizations. The emission line is a calibration line. Lower part: Energy level diagram corresponding to the polarized transition between two magnetic field split levels in $ErCrO_3$; this diagram corresponds to the Ia transition [CoHü75].

both polarizations, however, two pairs of lines are observed, where each pair is about 100 cm^{-1} from the other and the energy separation within the pairs is different in the two polarizations. One can interpret the observed splitting pattern as transitions between magnetic field split crystal field levels, where the origin for the difference in the splittings for the two polarizations is evident from the energy level diagram below the spectra. The sum of the splitting of one pair of lines in the two different polarizations gives twice the ground state splitting, and the difference twice the excited state splitting. The ground state splitting is much larger than the excited state splitting. The interpretation is substantiated by the intensities of the lines in each pair: The higher energetic component in each pair is the more intense one because it comes from the lower two of the internal magnetic field split ground state.

This experiment shows that the situation in the magnetically ordered state can indeed be treated in the same way as if an externally applied field were acting. Figure 29 shows the temperature dependence of the

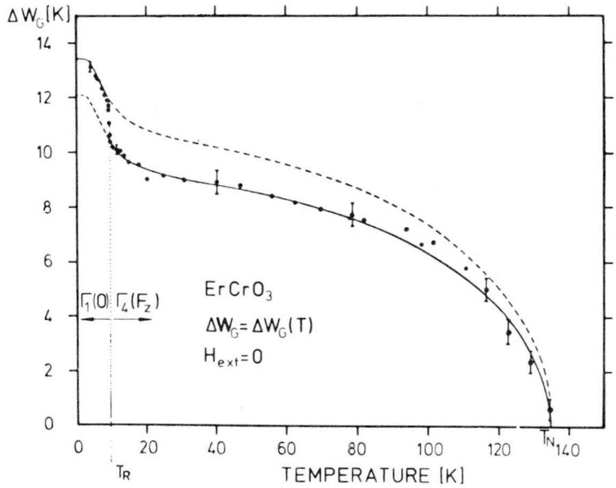

Fig. 29. Temperature dependence of the total ground state splitting ΔW_G of the erbium ions in $ErCrO_3$ as deduced from the $^4I_{15/2} \rightarrow {}^4S_{3/2}$ transition [CoHü75].

The dashed line gives the temperature dependence if no phase change were to take place, at the reorientation temperature T_R, meaning that in lowering the temperature, the Γ_4 phase would be stable down to 0 K, or in starting at 0 K, the Γ_1 phase would be stable up to $T = T_N$.

total ground state splitting in the magnetic order temperature range. For $T > 20$ K this curve (as expected) follows approximately a magnetization curve, the magnetization curve of the Cr sublattices. At low temperatures ($T < 20$ K) the interaction between the Er ions leads to an additional increase in the ground state splitting W_G. The sudden increase in W_G at $T_R = 9.4$ K is due to the Cr spin reorientation. The temperature dependence of W_G is that of the energy of an Er ion in $ErCrO_3$, which cannot be measured this easily by any other method. The dashed curves in the figure give the temperature dependence of W_G if for $T > T_R$ the Γ_1 phase would exist, and for $T < T_R$ the Γ_4 phase.

The information obtained from the optical spectra can be used to derive macroscopic magnetic properties. The susceptibility, magnetization, and specific heat calculated from the spectroscopic data are in good agreement with those measured directly. It is worthwhile to compare some results of optical spectroscopic measurements with those obtained by other techniques. To that purpose in Table 34 the ground state splitting, magnetic moment, and reorientation temperature obtained by different measuring techniques are presented. It can be seen that the ground state splitting and the magnetic moment ($\frac{1}{2}g_z$-factor) are best obtained by the

TABLE 34

MAGNETIC PARAMETERS FOR $ErCrO_3$ OBTAINED FROM OPTICAL EXPERIMENTS AND VARIOUS OTHER SOURCES[a]

		W_G (K)[b]			
Method	$T_R(0)$ (K)	$T < T_R$	$T > T_R$	$\frac{1}{2}g_z$ (μ_B)	Reference
Neutron diffraction	16.8			5.2 (4.2 K)	[BeMa67]
Specific heat	9.82 ($\pm$0.02)	13.5 Constant			[EiHo70]
Mössbauer		15.0 (4.2)	10.0 (10 K)	6.2 (4.2 K)	[EiCo69]
Specific heat, magnetization	9.0 ($\pm$1.0)	12.0 Constant		6.1 (1.5 K)	[VeAy70]
Optical spectroscopy	9.0 ($\pm$1.0)	13.2 (4.2 K)	10.1 (10 K)	5.8 ($\pm$0.4)	[MeMo70]
Optical spectroscopy	9.4 ($\pm$0.5)	13.2 ($\pm$0.1) (4.2 K)	10.4 ($\pm$0.2) (10 K)	5.68 ($\pm$0.2)	[CoHü75]
Optical spectroscopy	9.3 ($\pm$0.5)	13.4 ($\pm$0.6) (0 K)	10.3 ($\pm$0.2) (9.3 K)	5.65 ($\pm$0.15)	[HaHo75]

[a] From [CoHü75].

[b] W_G is the total splitting of the lowest crystal field level of the $^4I_{15/2}$ ground term by the internal fields acting on the Er ion.

optical technique, whereas a property like the phase transition temperature T_R can be determined with much greater accuracy by a specific heat experiment.

In this section we also have to discuss the first and perhaps still most noteworthy optical experiment on a magnetic RE material: Wickersheim's study [WiWh60, Wi61] of Yb–iron garnet (YbIG). This study showed for the first time the large anisotropy possible in the exchange interaction for RE ions, a consequence of the large orbital angular momentum of the RE wavefunctions.

The RE iron garnets are mixtures of iron oxide and RE oxides: $3(RE)_2O_3 \cdot 5Fe_2O_3$ (they are the magnetic analogs of the gallium and aluminum garnets; see Chapter 3). Their structure is cubic, but the point symmetry at the RE site is D_2. The crystal structure shows (and this has been confirmed by ESR experiments on nonmagnetic garnets) that there are six magnetically inequivalent RE sites (see Fig. 10). The Néel temperature of RE iron garnets is of the order of 500 K, but the investigations of

Pauthenet [Pa58, NePa64] show that the low temperature magnetic properties of these oxides are dominated by the magnetic moments of the RE ions. To understand their low temperature magnetic properties, a determination of their energy level diagram is desirable. This is difficult because of the low symmetry at the RE site and the large number of magnetically inequivalent sites. Therefore, although considerable work has been done on these compounds, reliable experimental data are available only for YbIG [Wi61] and ErIG [OrHü70]. Because the exchange interaction is now anisotropic, we can no longer describe it, as we did in eq. (6.7), by one quantity, the exchange field H^{ex}. In theory, however, it is still possible for the anisotropy in the exchange interaction to be the same as that of the g-tensor, in which case the interaction would still be described by the exchange field H^{ex}. As we shall see, this is not the case in practice. Thus the exchange interaction of every crystal field level has to be described by an interaction tensor. For this description it has become customary to use not the exchange field with its three principal values, but the energy splittings (called the G-tensor) produced by the interaction.

In spin Hamiltonian notation the splitting of a crystal field level in an external magnetic field H_{ext} by the exchange interaction can then be written

$$\Delta W = (\mu_B g H_{ext} + G)\bar{S} \tag{6.9}$$

To analyze the experimental data in terms of an interaction, we have the Hamiltonian of eq. (6.8). It can be parametrized in a way similar to the Hamiltonian for the crystal field interaction [Le64, Le66a], and the exchange interaction experienced by the RE ions can then be described by a number of exchange parameters (which have the same conceptual meaning as the crystal field parameters, namely, that they are primarily a way to catalog the experimental data; their a priori determination from the charge and spin distribution in the crystal, however, has not yet been achieved). In the garnet structure there are ten exchange constants necessary for a complete description of the RE iron exchange interaction. Note then that in order to determine all the exchange constants we need at least the exchange splittings in the three principal crystallographic directions of four crystal field levels.

The properties we are interested in order to determine the exchange constants are the principal values of the exchange tensor G for as many energy levels as possible. To determine G from the experimentally observed splitting, the splitting in the external magnetic field H_{ext} has to be negligible. In his experiment Wickersheim [Wi61] used an H_{ext} of 1.5 kOe, which was sufficient to saturate the YbIG crystal, but which produced no noticeable magnetic splitting of itself. The experiment consisted of rotating a plate of YbIG, cut so that it contained a (110) plane, in the magnetic field, which fixed the magnetization always parallel to its direction. A (110) plane includes at most four magnetically inequivalent sites (see Fig. 10). In the [100] direction we measure G_z (and g_z), and in the [110] direction G_x and G_y (g_x and g_y), as can be seen from Fig. 10. Figure 30 shows the exchange splitting in the (110) plane of the lowest crystal field level of the ${}^2F_{7/2}$ ground term and the ${}^2F_{5/2}$ excited term as obtained by Wickersheim [Wi61]. Table 35 gives the experimental G values obtained from the data in Fig. 30, along with the g-tensors for the two crystal field states. A comparison of the ratios of the g-tensor components and the G-tensor components shows that they are not linearly related. This is a clear manifestation of the anisotropy of the exchange interaction. Similar studies have been performed with ErIG [OrHü70], but in neither case are the experimental data sufficient for an unambiguous determination of the exchange parameters (in terms of eq. (6.8)). Therefore, a microscopic analysis of the magnetic properties of the RE iron garnets must await further experimental investigation [Le66a, Le69].

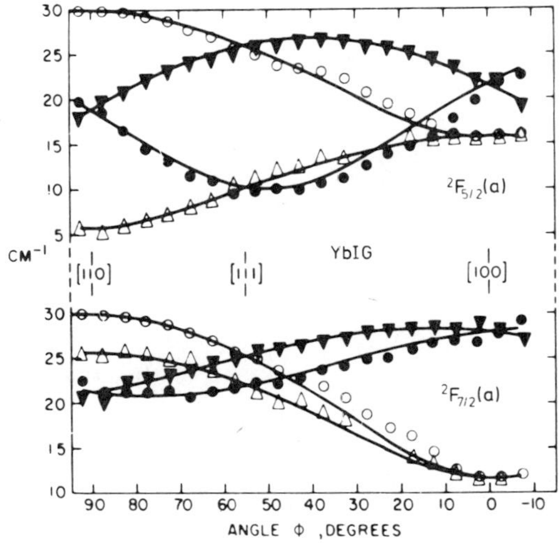

Fig. 30. Magnetic splitting of the $I^2F_{7/2} \rightarrow a^2F_{5/2}$ transition of YbIG in a (110) plane, $T = 77$ K, $H_{ext} = 1.5$ kOe [Wi61].

The angle ϕ is counted from the [100] direction. The [100], [111], and [110] directions are indicated.

TABLE 35

COMPONENTS OF THE EXCHANGE G- TENSOR IN YbIG AND OF THE g-TENSOR IN YbGaG FOR THE LOWEST CRYSTAL FIELD LEVELS OF THE $^2F_{7/2}$ AND $^2F_{5/2}$ TERMS OF Yb[a]

State	Principal value	g values: YbIG	g values: YGaG	G values
Lowest $^2F_{7/2}$ doublet	x	3.7	3.78	30.6
	y	3.6	3.60	26.1
	z	2.8	2.85	11.8
Lowest $^2F_{5/2}$ doublet	x	1.8	1.80	29.8
	y	1.8	1.98	5.6
	z	≤ 0.4	0.35	15.2

[a] From [Wi61] [WiWh62] [Le66].

6.5 Neighbor Interactions: $TbAlO_3$, $NdCl_3$, $DyVO_4$

In the preceding section the discrete nature of the crystal was neglected by making the assumption that the effect of the magnetic moments of the ions in the crystal can be approximated by a molecular field. Although this assumption is useful for interpreting many experimental facts, it is not entirely satisfactory because we know that a crystal is made up of distinct ions, each with a magnetic moment, and we are interested in the individual interactions. So far it has been possible to determine these individual interactions directly in one class of substances, those in which the magnetic interactions are to a good approximation of the Ising type ($J_{1z} \cdot J_{2z}$). That Ising-type interactions give rise to an easily analyzable structure in the energy level diagram was first realized by Bleaney *et al.* [BlEl51] in the ESR spectra of Nd^{3+}:$La(C_2H_5SO_4)_3 \cdot 9H_2O$. They observed, for the magnetic field parallel to the c axis of the crystal, two extra lines, one higher and the other lower in energy than the main line, which they interpreted in terms of the interaction with the two nearest neighbors of a particular ion: Depending on the orientation of the magnetic moments of such an ion triplet, three energy states are possible. The Ising interaction has an additional advantage. The change from one spin configuration to another is produced in first order by off-diagonal terms of the spin operator, that is, by terms of the form $J_{1+} \cdot J_{2-}$. For an exact Ising state, however, these terms are zero, and therefore the individual spin configurations are sufficiently long-lived to be observable.

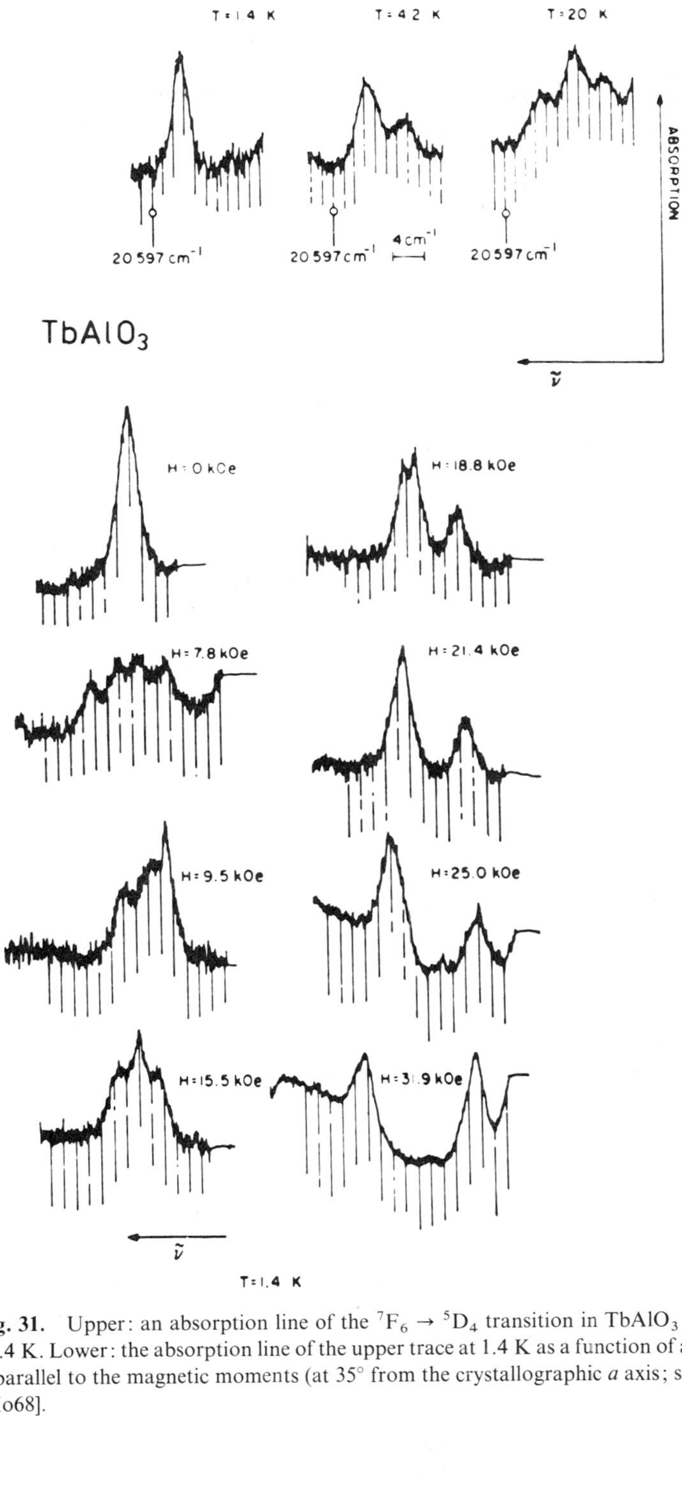

Fig. 31. Upper: an absorption line of the $^7F_6 \rightarrow {}^5D_4$ transition in $TbAlO_3$ at 20, 4.2, and 1.4 K. Lower: the absorption line of the upper trace at 1.4 K as a function of a magnetic field parallel to the magnetic moments (at 35° from the crystallographic *a* axis; see Fig. 32) [HüHo68].

In optical spectra of $NdCl_3$ and $NdBr_3$, Prinz [Pr66] was the first to observe structures that could be interpreted in terms of Ising interactions with two nearest neighbors. A characteristic of Nd salts is that all the energy levels are doubly degenerate, which, while providing additional information, also obscures the spectral phenomena somewhat. Thus we present the results for $TbAlO_3$, which has an accidental doublet as the ground state, whereas most of the excited states are singlets, which enables us to investigate the ground state properties via a single optical line. Figure 31 shows, at the top, one absorption line of the $^7F_6 \rightarrow {}^5D_4$ transition for $TbAlO_3$; at 20 K, we see the triplet structure expected for only near neighbor interactions, where the spectrum even shows roughly the correct intensity ratio between the first extra line, the main line, and the second extra line, namely 1:2:1; the Néel temperature of $TbAlO_3$ is $T_N = 3.95$ K [HoSh68], and 20 K is thus a temperature far above T_N. The observed structure in the absorption spectrum demonstrates that the neighbor configurations have lifetimes larger than the inverse of the linewidth observed (~ 2 cm^{-1}). At 4.2 K the thermal population changes the intensities of the various lines, and at 1.4 K only one line is observable, due to the depopulation of the higher states. Below T_N an additional shift of the line with the highest energy is observed: It reflects the long range interactions with ions other than the nearest neighbors.

Figure 32 shows a diagram of the unit cell [HüHo68], where the directions of the magnetic moments are also indicated [BiMa67]. The interaction above T_N is between ions of types 1 and 2 and of types 3 and 4

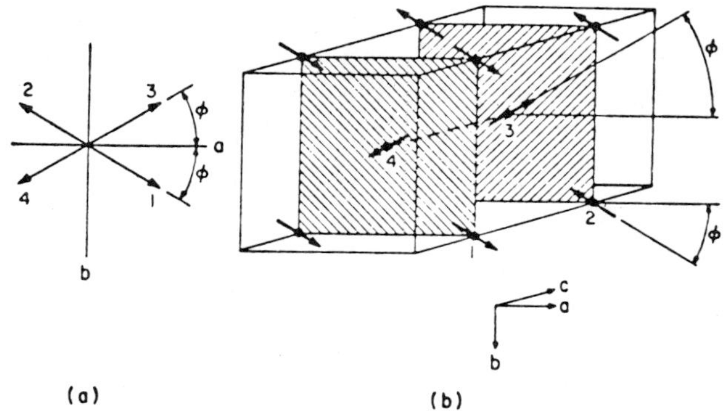

Fig. 32. Magnetic structure of $TbAlO_3$.
(a) Four-sublattice model deduced from magnetization data [HoSh68]; (b) magnetic unit cell as deduced from neutron diffraction data [BiMa67, HüHo68] (for the equations in Table 36, the magnetic field is along the direction of ion 1).

along the c axis. Since the splitting between two lines of a triplet at 20 K is 3.8 cm^{-1}, the interaction energy between two ions is 1.9 cm^{-1}. Of these, 0.7 cm^{-1} can be accounted for by dipolar interactions, leaving 1.2 cm^{-1} to the exchange interactions between two ions of types 1 and 2 or 3 and 4. For $T = 1.4$ K, below T_N, the total dipolar energy is 1.1 cm^{-1}; since an additional energy of 0.8 cm^{-1} is observed, a total of 1.1 cm^{-1} additional exchange energy must be present below T_N. (The total interaction is (3.8 ± 0.8) cm^{-1} and the exchange with the two ions along the c axis is 2.4 cm^{-1}.) It is reasonable to ascribe this additional energy to the interaction with the four neighbors in the a–b plane (ions of type 4 with respect to 1); thus the exchange interaction with each ion in the a–b plane is 0.25 cm^{-1} as compared to 1.2 cm^{-1} per ion with ions along the c axis.

With these values the behavior of $TbAlO_3$ in a magnetic field as seen from the absorption spectra will be examined. The direction and magnitude of the magnetic moment were determined from a Zeeman effect measurement: The moment was found to be $\mu = (9.0 \pm 0.5)\mu_B$, with an angle $\phi = \pm 35°$ with respect to the crystallographic a axis. If the antiferromagnetic structure shown in Fig. 32 is assumed, application of an external magnetic field in the direction of, e.g., ion 1 can produce two metamagnetic transitions at which the directions of the magnetic moments in a particular sublattice are reversed: one at a lower field H_{c1} when sublattice 2 changes its direction, and another at H_{c2}, when sublattice 4 changes its direction. From the fact that the moments in sublattices 1 and 4 have an angle of $2\phi = 70°$, we can predict that the ratios of the two fields should be roughly cos 70. If the foregoing description is correct, we expect qualitatively: for $H_{ext} < H_{c1}$,† four absorption lines for each transition (all four sublattices are unequal); for $H_{c1} < H_{ext} < H_{c2}$, three transitions (1 and 2 are equivalent but inequivalent to 3 and 4); and for $H_{c2} < H_{ext}$, two lines (1 and 2 are inequivalent to 3 and 4). That this is indeed observed can be seen in Fig. 31 where spectra in the range $0 < H_{ext} < 32$ kOe are shown for $T = 1.4$ K. Regions with four, three, and two absorption lines are evident. In the more careful interpretation that we shall give now, we shall see that matters are not quite so simple. Using the calculated dipolar fields and the exchange fields measured in the zero field experiments, we can calculate the Zeeman patterns. To that purpose we describe each interaction by a magnetic field H_{ij}, which is the magnetic field parallel to the

† The symbol H usually is used for the external magnetic field; in cases where a number of different magnetic fields have to be introduced, we shall write H_{ext} for the external applied magnetic field in order to distinguish it clearly from the other fields.

sublattice magnetization on sublattice i due to sublattice j. Every field is split up into an exchange (ex) and a dipole (dd) field; so we have

$$H_{ij} = H_{ij}^{ex} + H_{ij}^{dd}$$

where it is assumed that only H_{12}^{ex} and H_{14}^{ex} are different from zero. The total field acting on ion i in the antiferromagnetic state can be written

$$H_i^{af} = H^{af} = \sum_{j=1}^{4} H_{ij}$$

where j labels the sublattices. Using calculated dipole fields and the exchange splittings given earlier, we find (with $\mu = 9\mu_B$)

$$H_{11} = 2.6 \text{ kOe}, H_{12} = 6.9 \text{ kOe}, H_{13} = 0.3 \text{ kOe},$$

$$H_{14} = 0.9 \text{ kOe}, H^{af} = 10.7 \text{ kOe}.$$

Before writing the energies of the ions in the various sublattices, we review the symmetry relations

$$H_{ij} = H_{ji}; \qquad H_{11} = H_{22} = H_{33} = H_{44}$$

$$H_{12} = H_{34}; \qquad H_{14} = H_{23}; \qquad H_{24} = H_{13}$$

With these expressions we can write the energy expressions for the ions in the various sublattices if the external field is parallel to the magnetic moment of ion 1; these equations are given in Table 36. Figure 33 gives a comparison of the calculations using the equation in Table 36 (solid lines) and the experimental results (points). In each of the three regions ((a), (a–b), (b)) a schematic picture of the magnetic structure, where the four moments have all been put into the origin, are shown, together with the direction of the magnetic field H_{ext}. There is quite good agreement

TABLE 36

ENERGY SHIFTS FOR THE VARIOUS MAGNETICALLY INEQUIVALENT IONS FOR A MAGNETIC FIELD ALONG THE DIRECTION OF ION 1[a]

	(a)	(a–b)	(b)
Ion 1	$-\mu H$	$-\mu H + 2\mu H_{12}$	$-\mu H + 2\mu(H_{12} + H_{14})$
Ion 2	$+\mu H$	$-\mu H + 2\mu(H^{af} - H_{11})$	$-\mu H + 2\mu(H_{12} + H_{14})$
Ion 3	$-\mu H \cos 2\phi$	$-\mu H \cos 2\phi + 2\mu H_{14}$	$-\mu H \cos 2\phi + 2\mu(H_{12} + H_{14})$
Ion 4	$+\mu H \cos 2\phi$	$+\mu H \cos 2\phi + 2\mu H_{13}$	$-\mu H \cos 2\phi + 2\mu(H_{12} + H_{14})$

[a] From [HüHo68].

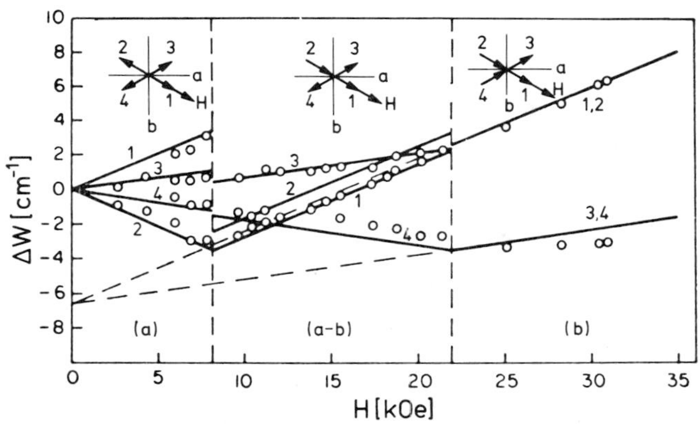

Fig. 33. Zeeman pattern of a typical absorption line of $TbAlO_3$ at 4.2 K. The magnetic field is directed at an angle of 35° from the *a* axis in the *a–b* plane. The sample was disk shaped with a demagnetization factor <0.001. The solid lines are calculated with the equations in Table 36 [HüHo68].

between the calculated curves and the data points. The most important difference between the qualitative discussion above and the exact calculated curves in Fig. 33 is a splitting of the curves for moments 1 and 2 in the region (*a–b*). This splitting is caused by taking into account interactions other than those between ions 1 and 2 and ions 3 and 4. (We see that if only the latter interactions are used, we have $H^{af} = H_{12}$, $H_{11} = H_{13} = H_{14} = 0$, and then indeed the curves for ions 1 and 2 coincide in region (*a–b*).) The splitting is, however, small ($\sim 1.2\ \mathrm{cm}^{-1}$) and therefore cannot be detected with the linewidth of $2\ \mathrm{cm}^{-1}$; this leads to the observed three-line pattern in region (*a–b*). From the equations in Table 36 we obtain for the two critical fields

$$H_{c1} = H^{af} - H_{11} = (8.1 \pm 2.1)\mathrm{kOe}$$

$$H_{c2} = (H_{c1} - 2H_{13})/\cos 70 = (21.9 \pm 6.1)\mathrm{kOe}$$

in reasonable agreement with the experimental results. Since $2H_{13} = 0.6$ kOe, the ratio of the two critical fields is indeed $\approx \cos 70$, as deduced from the previous simple arguments.

These detailed considerations are meant to show that the optical experiments can indeed produce microscopic data, which in turn can be used to interpret macroscopic measurements. A number of investigations of this type have been performed, with results of similar quality.

Let us go back to the experiment by Prinz [Pr66] on $NdCl_3$ and $NdBr_3$, which showed for the first time the direct observability of distinct neighbor interactions. Because all the states in these compounds are doublets, the neighbor interactions for a number of crystal field levels could be determined. They can be analyzed in terms of a spin Hamiltonian (effective spin $\bar{S} = \frac{1}{2}$)

$$\mathscr{H} = A_{ij}\bar{S}_{iz} \cdot \bar{S}_{jz} \tag{6.10}$$

with the results shown in Table 37. The magnetic dipolar contribution can be subtracted from the total observed coupling between neighboring ions.

The real exchange constants can then be extracted from the coupling constants, as defined by the effective spin $\bar{S}$, by dividing out the splitting factor of the state and by projecting the real spin S onto the total angular momentum J. This means that if the real spin Hamiltonian is

$$\mathscr{H} = J_{ij}S_{iz}S_{jz} \tag{6.11}$$

we have

$$A_{ij} = (g_{Ji}g_{Jj})^{-1}(g_{Ji} - 1)(g_{Jj} - 1)g_{zi}g_{zj}J_{ij} \tag{6.12}$$

where g_{zi} and g_{zj} are the g-factors (effective spin $\bar{S} = \frac{1}{2}$) parallel to the z axis and g_{Ji}, g_{Jj} are the Landé g_J-factors; the exchange constants J_{ij} determined by this procedure are shown in Table 37, where we see large

TABLE 37

NEIGHBOR INTERACTIONS IN $NdCl_3$ AND $NdBr_3$ FOR VARIOUS TERMS[a]

Crystal field state	Term	A_{ij} (cm^{-1})	$\frac{(g_J - 1)}{g_J}$	J_{ij} (cm^{-1})
$NdCl_3$				
I[b]	$^4I_{9/2}$	1.62	−0.375	0.60
b	$^4F_{9/2}$	1.06	0.250	0.32
d	$^4F_{9/2}$	−0.28	0.250	−0.50
a	$^4G_{9/2}$	0.02	0.147	0.07
b	$^2D_{5/2}$	0.20	0.166	0.54
$NdBr_3$				
I	$^4I_{9/2}$	2.22	−0.375	1.19

[a] After G. A. Prinz, *Phys. Rev.* **152**, 474 (1966).

[b] The crystal field states are labeled with roman numerals for the ground term and with lowercase letters for the excited terms, each starting at the state with the lowest energy.

differences between the various crystal field levels. The exchange integrals in Table 37 also show that although most crystal field levels have an antiferromagnetic exchange coupling (positive sign in the sign convention chosen), the coupling constant is ferromagnetic for one crystal field level. This shows that the exchange coupling depends strongly on the wave-functions of the particular crystal field state.

Effects on optical spectra of interactions with distinct neighbors have also been found in some RE zircons, notably $DyPO_4$ [WrMo71b] and $DyVO_4$ [WrMo71a, KaBe74]. In $DyPO_4$ and $DyVO_4$ the interactions are of the Ising type and each ion interacts predominantly with its four nearest neighbors. The various orientation possibilities in the neighbor configurations of an ion give rise to five different energy levels for each ion, resulting in an experimentally observed quintuplet structure of the absorption lines in these compounds. $DyPO_4$ and $DyVO_4$ are very well-behaved three-dimensional Ising systems, and by using optical data in conjunction with the specific heat and magnetization measurements a very consistent explanation of their magnetic properties can be achieved [WrMo71b].

In other zircons, like $HoPO_4$ and $HoAsO_4$, which are also three-dimensional Ising systems, the interactions are of much longer range [LaKa75, SchwBe74] and therefore neighbor interactions do not show up explicitly in the spectra. In these cases a determination of the interactions of the various neighbor shells can only be obtained by the use of specific heat and magnetic data. Table 38 shows the results of such analysis, which made it possible to obtain the interactions with three neighbor shells. The interactions are expressed as magnetic fields, where

TABLE 38

EFFECTIVE FIELDS IN THE ANTIFERROMAGNETIC STATE FOR $HoPO_4$ AND $HoAsO_4$[a]

Order of neighbor	Number of equivalent neighbors z	$HoPO_4$ H^{ex} (kOe)	$HoPO_4$ H^{dd} (kOe)	$HoPO_4$ H^{af} (kOe)	$HoAsO_4$ H^{ex} (kOe)	$HoAsO_4$ H^{dd} (kOe)	$HoAsO_4$ H^{af} (kOe)
nn	4	−1.10	+3.00	+1.90	−1.56	+2.47	+0.91
nnn	8	−0.35	−0.55	−0.90	−0.70	−0.39	−1.09
tnn	4	+3.50	−1.50	+2.00	+2.16	−1.29	+0.87
Long range			+1.08	+1.08		+0.86	+0.86
Total		+2.05	+2.03	+4.08	−0.10	+1.65	+1.55

[a] From [LaKa75]. A positive sign means the field is parallel to the magnetic moment of the central ion.

again the total field H^{af} has been broken down into a dipolar and an exchange contribution. For an interpretation of the results we have to note that under the assumed magnetic structure [SchwBe74] the nn ions are antiparallel to the central ion, the nnn ions are parallel, and the tnn (third nearest neighbour) ions are again antiparallel (two sublattice antiferromagnet). It is surprising that in both holmium zircons the dominant antiferromagnetic exchange interaction is with the tnn ions, whereas the exchange interaction with the nn ions is ferromagnetic.

We now revert to our discussion of the metamagnetic behavior of an Ising antiferromagnet, because things are not quite so simple as we implied earlier. As will be shown in $DyVO_4$ [KaBe74], when a magnetic field is applied to an Ising antiferromagnet, for certain values of this magnetic field, a state is observed in which a saturated paramagnetic phase and an antiferromagnetic phase coexist, as first proposed by Wyatt [Wy68] and investigated in great detail in dysprosium aluminium garnet (DAG) by Landau *et al.* [LaKe71]. A diagram of the three different phases (antiferromagnetic, mixed, and saturated paramagnetic) is shown in Fig. 34. For the sake of argument, we assume from now on that the samples have an ellipsoidal shape with demagnetization factor N, and the temperature is $T < T_N$. If an external field H_{ext} is applied to the sample, two critical fields are observed, H_{c1} and $H_{c2} > H_{c1}$, which are associated with the transition from the antiferromagnetic to the mixed and from the mixed to the saturated paramagnetic phase. For a sample with $N = 0$, the two-phase transitions coincide, and this observation is the starting point for the discussion of the mixed-phase phenomenon.

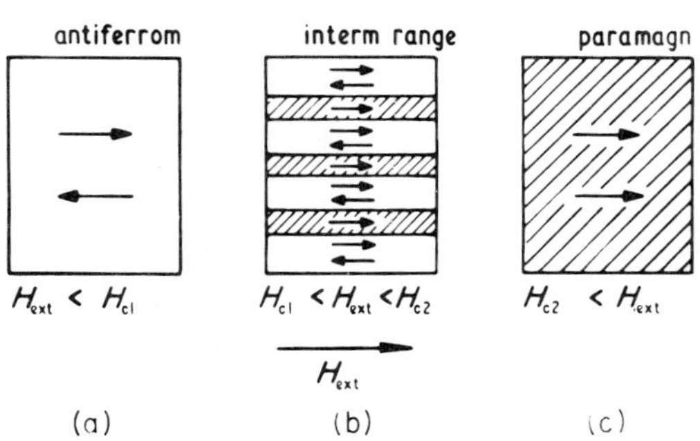

Fig. 34. The three phases in an antiferromagnet as suggested by Wyatt [Wy68, KaBe74] (Copyright by The Institute of Physics).

(a) Normal antiferromagnetic phase; (c) saturated paramagnetic phase. In the intermediate phase (b), thin sections of the saturated paramagnetic phase are intercalated into the antiferromagnetic phase.

For a needle-shaped sample with $N = 0$, the transition at $H_{c1} = H_{c2}$ is a metamagnetic transition from the antiferromagnetic phase to the saturated paramagnetic phase. For a sample with demagnetization factor $N > 0$ this cannot happen. If at $H_{ext} = H_{c1}$ all the spins were to flip, creating a saturation magnetization M_s, a demagnetization field $4\pi NM_s$ would occur in the sample, reducing the internal field to $H_{int} = H_{c1} - 4\pi NM_s$. Since for $N > 0$ this field is smaller than H_{c1}, the magnetized sample cannot exist. As a matter of fact, the phase transition can only be completely at a field $H_{ext} = H_{c2} = \mathrm{H}_{c1} + 4\pi NM_s$. In the field range $H_{c1} < H_{ext} < H_{c2}$ the sample is in the intermediate state, which may be visualized as a spatial mixture of needle-shaped saturated paramagnetic state regions that are intercalated into the antiferromagnetic regions. The volume of these needles will adjust itself so that $H_{int} = H_{ext} - 4\pi NM = H_{c1}$ always (M is the average magnetization of the sample). Only at $H_{ext} = H_{c1} + 4\pi NM_s = H_{c2}$ does $H_{int} = H_{c1}$ for the whole sample, and thus the sample is in the fully magnetized state.

Now the effect of the mixed phase on the optical spectrum is calculated; the internal field in the needle-shaped region is

$$\mathrm{n}: \quad H_{int} = H_{ext} + H^{\mathrm{n}} - 4\pi mNM_s$$

where H^{n} is the exchange field of the nearest neighbors and the dipole field of the needle; m is the reduced magnetization M/M_s.

Outside the needles in the antiferromagnetic regions there are two possibilities. One has to determine the field at the site of ions that have their magnetic moment parallel to the external field (af(+)) and for those that have it antiparallel to the external field (af(−)). Thus we have

$$\begin{aligned} \mathrm{af}(+)&: \quad H_{int} = H_{ext} + H^{\mathrm{af}} - 4\pi mNM_s \\ \mathrm{af}(-)&: \quad H_{int} = -H_{ext} + \mathrm{H}^{\mathrm{af}} + 4\pi mNM_s \end{aligned} \tag{6.13}$$

where the second terms are the effective fields parallel to an antiferromagnetically ordered ion.

By minimizing the total energy of the sample we find

$$m = \frac{-2H_{ext} + H^{\mathrm{af}} - H^{\mathrm{n}}}{-8\pi NM_s} \tag{6.14}$$

showing that the magnetization is a linear function of the external field.

For $m = 0$ and $m = 1$ we find

$$\begin{aligned} H_{c1} &= \tfrac{1}{2}(H^{\mathrm{af}} - H^{\mathrm{n}}) \\ H_{c2} &= H_{c1} + 4\pi NM_s \end{aligned} \tag{6.15}$$

From this

$$\begin{aligned} \text{n:}\quad & H_{\text{int}} = \tfrac{1}{2}(H^{\text{af}} + H^{\text{n}}) \\ \text{af}(+)\text{:}\quad & H_{\text{int}} = \tfrac{3}{2}(H^{\text{af}} - \tfrac{1}{2}H^{\text{n}}) \\ \text{af}(-)\text{:}\quad & H_{\text{int}} = \tfrac{1}{2}(H^{\text{af}} + H^{\text{n}}) \end{aligned} \tag{6.16}$$

meaning that above H_{c1} we have two field independent lines. A calculation of the dipole fields and an analysis of the quintuplet separation and of the ground state splitting for $T \to 0$ K yields $H^{\text{n}} = 3.9$ kOe and $H^{\text{af}} = 8.1$ kOe; thus

$$\begin{aligned} \text{n:}\quad & H_{\text{int}} = 6.0\ \text{kOe} \\ \text{af}(+)\text{:}\quad & H_{\text{int}} = 10.2\ \text{kOe} \\ \text{af}(-)\text{:}\quad & H_{\text{int}} = 6.0\ \text{kOe} \end{aligned} \tag{6.17}$$

Figure 35 shows a plot of the magnetic field dependence of the 20962.5 cm^{-1} absorption line in $DyVO_4$ and the two critical fields H_{c1} and H_{c2}, and the field independent behavior between them can be seen.

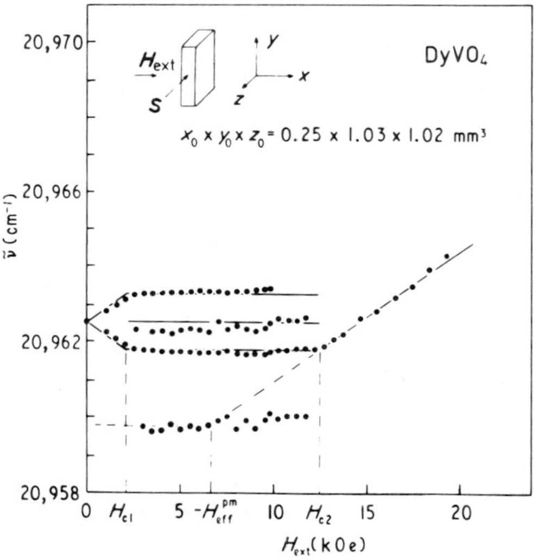

Fig. 35. Zeeman pattern of the $I^6H_{15/2} \to a^4F_{9/2}$ transition in $DyVO_4$ at $T = 1.4$ K in light polarized parallel to the applied field H_{ext} [KaBe74] (Copyright by The Institute of Physics).

Note that the magnetic field is applied to the sample in such a way that demagnetization effects are large. $H^{\text{pm}}_{\text{eff}}$ is the sum of the exchange and dipole fields experienced by an ion in the saturated paramagnetic state. **S** is the propagation direction of the light. H_{c1} and H_{c2} are the two critical fields between which the intermediate phase exists. x, y, z are the unit cell directions.

For $H < H_{c1}$ a splitting of the zero field line into two lines is observed—corresponding to the ions on the sublattice whose moments are parallel and antiparallel to H_{ext}, respectively; at $H = H_{c1}$ the mixed phase starts to occur and from this field up to H_{c2} the line positions are field independent. For $H > H_{c2}$ only one sort of magnetic ion is present, and only one absorption line is seen.

With the demagnetization factor of the sample that was used to obtain the data in Fig. 35 we find

$$H_{c1} = 2.1 \text{ kOe}; \qquad H_{c2} = 13.1 \text{ kOe} \tag{6.18}$$

in good agreement with the results in Fig. 35.

The ions in the surface between the two regions give rise to an additional line (20,962.6 cm^{-1}) between the two just calculated for $H_{c1} \leq H_{ext} \leq H_{c2}$. That this line is visible indicates that the needles are very thin and numerous, producing a great many surface ions. If in the extreme the needles are assumed to consist only of strings of ferromagnetically coupled ions, another type of site is also possible, namely, those at the ends of the strings. This may explain the extra line, also field independent, found at lower energy (20,959.8 cm^{-1}) in Fig. 35.

Again for this phenomenon optical measurement may yield useful answers to specific questions even though other measurements were needed to detect the phenomenon in the first place. The question now arises whether it is wrong to neglect the mixed phase when discussing $TbAlO_3$. It is not, because the range of external field in which the mixed phase exists is determined by the size of N, the demagnetization factor. Optical absorption experiments often involve very thin plates, which have quite a small N; thus the mixed phase range is small and the simplified treatment given for $TbAlO_3$ is approximately correct. The wide field range for the mixed phase found in the $DyVO_4$ (Fig. 35) was obtained by using a specially shaped sample that had a large demagnetization factor.

The discussion in this section of the observability of neighbor effects has implicitly introduced the time domain: For a configuration of ordered spins to be observable, the configuration lifetime must be longer than the time it takes to make the measurement, which in an optical experiment is the lifetime of the excited optical state (10^{-10} sec or less). An optical experiment can therefore measure the short range order in a magnetic system, which also exists above the transition temperature. This short range order gives the magnetic internal energy of the system, which is obtained by integrating the specific heat.

This phenomenon has been investigated in detail in dysprosium aluminium garnet (DAG) [CoGe65, Fa69, GeLe69], a three-dimensional Ising antiferromagnet. The Ising nature of the interactions again ensures that the magnetic clusters are relatively long-lived, which makes this effect observable by optical means. Figure 36 summarizes the results of neutron diffraction, specific heat, and optical absorption experiments, together with a Monte Carlo method calculation of the sublattice magnetization and the internal energy. All curves have been normalized. The neutron diffraction experiment measures the sublattice magnetization, the long range order, which vanishes at $T = T_N$. The internal energy from a specific heat experiment measures above T_N the short range order. It extends over T_N and is in good agreement with the normalized line shift measured by an optical absorption experiment, showing that here the optical experiment measures the short and not the long range order. The short range order can be described by the spin correlation function. The Monte Carlo calculation is quite successful in explaining the various experimental data. In this case the interaction is also one with distinct neighbor configuration, namely that produced by the short range order, but the number of neighbors is so large that the individual contributions cannot be resolved.

We have dealt with the different cases in some detail to show that care must be exercised in analyzing optical experiments with the aim of obtaining magnetic data from them. It is not obvious beforehand whether an

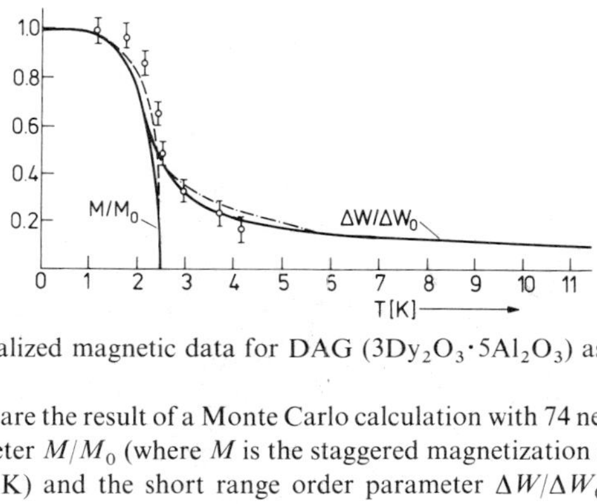

Fig. 36. Normalized magnetic data for DAG ($3Dy_2O_3 \cdot 5Al_2O_3$) as a function of temperature [Fa69].

The solid curves are the result of a Monte Carlo calculation with 74 neighbors for the long range order parameter M/M_0 (where M is the staggered magnetization at $T > 0$ K and M_0 its value at $T = 0$ K) and the short range order parameter $\Delta W/\Delta W_0$ (where ΔW is the magnetic splitting of the lowest crystal field level of the ground term at $T > 0$ K, and ΔW_0 its value at $T = 0$ K). –·–·– represents specific heat; –––, neutron diffraction; ϕ, optical absorption ${}^6H_{15/2}(I) \rightarrow {}^4F_{9/2}(b)$. The optical data points and the integrated magnetic specific heat parallel the short range order parameter, showing that they are a measure of this parameter.

optical experiment measures the long range or the short range parameter, which has to be determined by careful experimentation.

6.6 Spin-Wave Sidebands: $GdCl_3$

This type of interaction is the domain of the 3d-transition metal compounds and has seldom been observed with REs. Two experiments along this line warrant mention. First, Meltzer *et al.* [MeCo68, MeMo72] investigated the spectra of $GdCl_3$ ($T_c = 2.2$ K) and $Gd(OH)_3$ ($T_c = 0.94$ K) in strong magnetic fields and obtained some information about the spin-wave energies in these compounds. Second, Belorizky *et al.* [BeLe70] forced erbium aluminium garnet with fields up to 90 kOe into a ferromagnetic state. They then observed optical transitions that are analogous to the spin-wave sidebands. Similar effects were observed by Becker *et al.* [BeLe72a] in $TmVO_4$ and by Becker and Laugsch [BeLa73] in $TmAsO_4$.

In order to show the shortcomings and possibilities of these types of experiments, the results of Meltzer and Moos [MeMo72] will be outlined briefly. The ferromagnetic transition temperature of $GdCl_3$ is so low that the magnetic interactions give splittings and energy level shifts that cannot be seen clearly even at very high resolution. Therefore, a large magnetic field (35 kOe) was applied in order to separate the various energy levels sufficiently. This makes the experimental situation very similar to one where a large magnetic field is applied in the paramagnetic state.

$GdCl_3$ is a hexagonal crystal. Each Gd^{3+} ion has two nearest neighbors along the crystal axis ($d_{nn} = 4.105$ Å) and six next nearest neighbors ($d_{nnn} = 4.73$ Å); the other neighbors are much farther away, so interaction with them can be neglected. Ground state measurements have yielded interaction parameters with these two types of neighbors of $J_{nn} = +0.054\ \mathrm{cm}^{-1}$ and $J_{nnn} = -0.066\ \mathrm{cm}^{-1}$. Note that although the Gd ions order ferromagnetically, the interaction along the hexagonal axis is antiferromagnetic [HuBi68, ClWo68].

The description of the system closely follows that known and used successfully for 3d ions. As has been mentioned, the ground state of a 3d ion is a pure spin state; the spins interact with each other and in the magnetically ordered state a coherent motion of the spins in the sample takes place. The elementary excitations of this motion are called magnons or spin waves. The magnon system can be described by the Hamiltonian:

$$\mathscr{H} = + \sum_{ij} J_{ij}\mathbf{S}_i \cdot \mathbf{S}_j + (g\mu_B)^2 r_{ij}^{-3}[\mathbf{S}_i \cdot \mathbf{S}_j - 3(\mathbf{S}_i \cdot \mathbf{r}_{ij})(\mathbf{S}_j \cdot \mathbf{r}_{ij})/r_{ij}^2] + g\mu_B H \sum_i S_{iz} \tag{6.19}$$

where the crystal field interaction can be neglected because the $^8S_{7/2}$ ground term of Gd is an S state. Here i and j label the ions, and $\mathbf{r}_{ij}$, J_{ij}, and $\mathbf{S}$ are the distance, interaction parameter, and spin operator, respectively, of these ions. In eq. (6.19) the summation runs over all the ions in the crystal. The Hamiltonian in eq. (6.19) is used in a linear spin-wave approximation to calculate the dispersion relations for the ground state. Similarly, for the excited state a Hamiltonian of the form

$$\mathscr{H} = \sum_i \mathscr{H}_i + \sum_{i,j} V_{ij} \tag{6.20}$$

is introduced, where $\mathscr{H}_i$ is the single ion Hamiltonian containing the crystal field, the dipolar interactions with other ions, and the external magnetic field. V_{ij} is the Coulomb interaction between two ions where one ion is in the ground state and the other in an excited state. The dispersion relations are then obtained as in the magnon case.

Now a brief comment is necessary because the foregoing seems to be in contradiction to much that has been said up to here. So far in this book the dispersion of the RE energy levels has been neglected; that is, we have assumed that the W($\mathbf{k}$) relation, where $\hbar\mathbf{k}$ is the crystal momentum, is a horizontal function of $\mathbf{k}$. Of course, this is not correct, since the RE ions are situated in a crystal and their wavefunctions overlap slightly, so that their W($\mathbf{k}$) relations must have some dispersion (rough estimates show it to be well below 1 cm^{-1} for typical crystals; see, however, [CoMe75, Me76]). In a rigorous sense the RE states in crystals are therefore states of the whole crystal. This means that they are excitons (a general name for electronic crystal states) rather than single ion states. Due to the smallness of the dispersion, in almost every case the single ion state description is, however, fully appropriate, and we have used it throughout the book. However, when analyzing lineshapes in strongly interacting systems, as in the case of the spectra of $GdCl_3$ and $Gd(OH)_3$, we cannot neglect the ion–ion Coulomb interaction and the exciton formalism has to be used.

Experimentally, the transitions $^8S_{7/2} \rightarrow {}^6P_{5/2}, {}^6P_{7/2}$ were observed at 1.4 K in an external field of 35 kOe. The line shapes observed are compared with those calculated from the dispersion relations by performing temperature averaging and leaving the exciton interaction parameters (V_{ij}) as free variables. The magnon dispersion relations are calculated from the known ground state interaction parameters presented earlier. Figure 37 shows a comparison of measured line shapes and those calculated with the quoted magnon parameter and exciton parameters from Table 39 that gave the best fit to the experimental results.

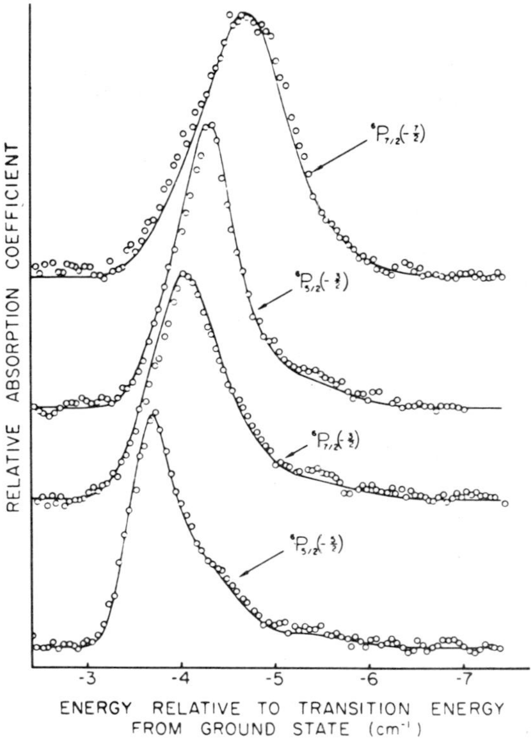

Fig. 37. Measured and calculated magnon–exciton transition line shapes in $GdCl_3$ ($^8S_{7/2} \to {}^6P_{5/2}, {}^6P_{7/2}$) [MeMo72], where $H = 34.87$ kG and $T = 1.40$ K.

The circles are observed data, the lines are calculated data. The energy is given with respect to the line position at $H = 0$. The states are characterized by their dominant M_J composition.

In Table 39 $\Delta\tilde{\nu}$ is a Gaussian-shaped linewidth that had to be given to each transition. The numbers in the table show that the interaction parameters for the excited states are quite small, although they are not negligible. It is interesting that exciton effects can indeed be observed, even if (as expected) they are small. Thus, this experiment serves to confirm the single-ion description of RE properties in insulating crystals. This means that the dispersion of the crystal field states can be neglected which is a result of the small ion–ion interaction.

Recent experiments of Cone and Meltzer [CoMe75, Me76] can, however, be used to constrain this statement considerably. They found, in one crystal field level of the $^7F_5$ term in $Tb(OH)_3$, a dispersion of

TABLE 39

INTERACTION PARAMETERS FOR $GdCl_3$ AS OBTAINED BY ANALYZING THE $^8S_{7/2} \rightarrow {}^6P_{5/2}$, $^6P_{7/2}$ LINE SHAPES[a]

State $^6P_J(M_J)$	V_{nn} (cm^{-1})	V_{nnn} (cm^{-1})	$\Delta\tilde{\nu}$ (cm^{-1})
$^6P_{7/2}(-7/2)$	0.10	−0.01	0.54
$^6P_{7/2}(-5/2)$	−0.05	0.005	0.27
$^6P_{7/2}(-3/2)$	0.005	−0.01	0.54
$^6P_{5/2}(-5/2)$	−0.04	−0.01	0.54
$^6P_{5/2}(-3/2)$	0.04	−0.01	0.40

[a] From [MeMo72]. $\Delta\tilde{\nu}$ is the half-width of the transitions. The magnon dispersion parameters are given in the text.

about 10 cm^{-1}, much larger than might have been anticipated. So far this seems to be the only well-documented case where such a large dispersion has been established and at present it is not known whether this result manifests a coincidence of some unusual circumstances, or whether the dispersion caused by the ion–ion interaction has escaped notice so far. It seems appropriate, however, to reinvestigate spectra of other crystals in order to make sure that a very interesting phenomenon and a source of important information on ion–ion interaction has not been overlooked.

CHAPTER 7

Hyperfine Interactions

Since the RE optical absorption lines are very narrow, we might ask whether they also permit detection of interactions with the nuclear moments (hyperfine structure). Except for one notable exception such detection is not possible by direct spectroscopy: The hyperfine interactions give energy level splittings of at most 1 cm^{-1} and this is about the residual linewidth due to strains in the crystal. However, the hyperfine structure of holmium in $Ho^{3+} : LaCl_3$ and $Ho^{3+} : Y(C_2H_5SO_4)_3 \cdot 9H_2O$ (see Fig. 2) has been observed and analyzed in detail. This observation is made possible by the fortunate coincidence of a very small residual linewidth, a very large magnetic nuclear moment, and the existence of only one isotope, ^{165}Ho [DiCr67, PeHü70].

Interest in the hyperfine interactions in Ho compounds was stimulated not so much by the need to measure, e.g., its nuclear moment as by the possibility of obtaining information about the interaction of the RE ion with the crystal field by an analysis of the quadrupole interaction.

The largest contribution to the hyperfine interaction is generally that of the nuclear magnetic moment; the quadrupole interaction is much smaller. The latter is the product of the nuclear quadrupole moment Q and the electrical field gradient (the second derivative of the electrical potential at the site of the nucleus) seen by it. For an RE ion sitting in a crystal, this electric field gradient has two contributions: one from an aspherical charge distribution of the 4f electrons and one from the aspherical charge distribution of the surrounding ions. Both contributions, however, are modified from their pure values by the fact that the closed shells of the RE ions are polarized by these electrical field gradients and can thus enhance or diminish them; this effect is called the Sternheimer [Ste51] effect. The quadrupole interaction is therefore written as

$$W_Q \sim Q\{(1 - R_Q)q_e + (1 - \gamma_\infty)q_{lat}\} \tag{7.1}$$

where q_e is the field gradient produced by 4f electrons at the nucleus, q_{lat} is the field gradient produced by the lattice at the nucleus, and $(1 - R_Q)$ and $(1 - \gamma_\infty)$ are the Sternheimer factors, which correct for the polarizability of the closed shells of the RE ion. These factors are of interest because they can be compared with those obtained from theoretical atomic wavefunctions [WaFr64a, FrWa63].

Since eq. (7.1) for the quadrupole interaction contains two contributions, evaluation of the two Sternheimer factors from the measured quadrupole interaction constant is not possible; two different measuring techniques have to be combined in order to evaluate these parameters. This can be done by using an optical experiment that measures only $(1 - R_Q)$, because such experiments disclose only $\Delta M_I = 0$ transitions (where M_I is the nuclear magnetic quantum number), and an ESR experiment which, via the $\Delta M_S = \pm 1$, $\Delta M_I = \pm 1$ transitions, measures both contributions. (The $\Delta M_S = \pm 1$, $\Delta M_I = 0$ transitions are the ones normally observed in EPR experiments; with extra care, however, those mentioned can be also observed.)

Our discussion will be restricted to the ethylsulfate and chloride of Ho. For axial symmetry the hyperfine interaction Hamiltonian can be written as [AbBl70]

$$\mathscr{H}_{hfs} = A_J \mathbf{I}\cdot\mathbf{J} + \{P^J_{4f}(J_z^2 - \tfrac{1}{3}J(J+1)) + P_{lat}\}\{I_z^2 - \tfrac{1}{3}I(I+1)\} - g_N \mu_B \mathbf{I}\cdot\mathbf{H}_{ext} \tag{7.2}$$

where

$$\begin{aligned} A_J &= \frac{2\mu_I \mu_N \mu_B}{I} N_J \langle r^{-3}\rangle_{4f}(1 - R_M) \\ P^J_{4f}(J_z^2 - \tfrac{1}{3}J(J+1)) &= \frac{3e^2 Q}{4I(2I-1)} q_{zz}^{4f}(1 - R_Q) \\ q_{zz}^{4f} &= -3\alpha_J \langle r^{-3}\rangle_{4f}(J_z^2 - \tfrac{1}{3}J(J+1)) \\ Q &= \langle I\ M_I{=}I|3z_n^2 - r_n^2|I\ M_I{=}I\rangle \\ P_{lat} &= \frac{3e^2 Q}{4I(2I-1)} q_{zz}^{lat}(1 - \gamma_\infty) \\ q_{zz}^{lat} &= -\frac{4A_{20}\langle r^2\rangle}{e^2\langle r^2\rangle_{4f}(1 - \sigma_2)} \end{aligned} \tag{7.3}$$

TABLE 40

SHIELDING FACTORS FOR SOME RARE EARTH COMPOUNDS[a]

(a) R_Q

RE ion	Compound	Method	Q $(10^{-24}\ cm^2)$[b]	R_Q	Reference
^{161}Dy	Atom	Atomic beam		0.35	[FeEb64]
^{163}Dy	Atom	Atomic beam	2.60	0.32	[FeEb64]
$^{161}Dy^{3+}$	Metal	Mössbauer	2.60	0.50	[OfRa65]
^{165}Ho	Atom	Atomic beam	3.53	0.31	[GoSc65]
$^{165}Ho^{3+}$	HoEs[g]	Optical spectroscopy	3.53	0.24	[PeHü70]
$^{165}Ho^{3+}$	$Ho^{3+}:LaCl_3$	Optical spectroscopy	3.53	0.24	[PeHü70]
^{167}Er	Atom	Atomic beam	3.65	0.29	[SmUn65]
$^{169}Tm^{3+}$	$TmFe_2$	Mössbauer	1.5	0.30	[Co64]
$^{169}Tm^{3+}$	TmES	Mössbauer	1.5	0.23	[MöBa64]
$^{169}Tm^{3+}$	Tm_2O_3	Mössbauer	1.5	0.13	[MöBa64]

(b) $\dfrac{1-\gamma_\infty}{1-\sigma_2}$

RE ion	Compound	Method	Q (barns)	$A_{20}\langle r^2\rangle$ (cm^{-1})[c]	$\dfrac{1-\gamma_\infty}{1-\sigma_2}$	Reference
$^{154}Eu^{3+}$	EuES	NO[e]	3.75	80	360(40)	[BlSh66]
$^{161}Dy^{3+}$	DyES	Mössbauer	2.6	124	250(30)	[WiNo66]
$^{165}Ho^{3+}$	HoES	ESR[f] + optical spectroscopy	3.53	143	240(80)	[PeHü70]
$^{167}Er^{3+}$	ErES	ESR	3.65	119	210(30)	[BoDu52]
$^{169}Tm^{3+}$	TmES	Mössbauer	1.5	130	250(30)	[MöBa64]
$^{165}Ho^{3+}$	$Ho^{3+}:LaCl_3$	ESR + optical spectroscopy	3.53	114	170(90)	[PeHü70]
$^{169}Tm^{3+}$	Tm_2O_3	Mössbauer	1.5	-82[d]	120(20)	[MöBa64]

[a] From [PeHü70].
[b] Coulomb excitation measurements by [OlEl60].
[c] $A_{20}\langle r^2\rangle = A_{20}(1-\sigma_2)\cdot\langle r^2\rangle$ (point charge).
[d] $A_{22}\langle r^2\rangle = -636\ cm^{-1}$.
[e] Nuclear orientation.
[f] Electron spin resonance.
[g] Es stands for ethylsulfate.

The symbol μ_I represents the nuclear moment, μ_N the nuclear magneton, N_J (often written $(J\|N\|J)$) the equivale·• operator for the hyperfine interaction [AbBl70], R_M the Sternheimer factor for the magnetic hyperfine interaction, r_n the radius vector of the nuclear coordinate, and z_n its z components; q_{zz} equals q_e if the ion is in a lattice of axial symmetry (as is the case for the ethylsulfates).

The crystal field parameter $A_{20}\langle r^2 \rangle$ is defined in Section 3.2. To obtain the field gradient from this crystal field parameter, we have to correct for the square of the radial distribution of the 4f electrons $\langle r^2 \rangle_{4f}$ and for the shielding of the field gradient by the closed shells when seen by the 4f electrons, which is done by the shielding factor $(1 - \sigma_2)$. The quadrupole interaction, as analyzed with the help of optically determined crystal field parameters, therefore gives two shielding parameters $(1 - R_Q)$ and $(1 - \gamma_\infty)/(1 - \sigma_2)$.

The optical experiments were performed by placing a Fabry–Pérot interferometer in front of the grating spectrograph in order to increase the resolution. A typical spectrum taken with such an arrangement was shown in Fig. 2, where the hyperfine splitting of the electronic lines is clearly seen.

Analysis of the measurements yielded the shielding constant values that are shown together with those for other RE elements, in Table 40. The numbers are reasonably consistent. The antishielding of the lattice field gradient by the closed shells is appreciable.

The linewidth problem can be circumvented by using a microwave/optical double resonance technique [HuLi76). In these experiments the intensity of a fluorescent line is monitored as the magnetic field is swept and a constant microwave frequency is applied. As the various hyperfine transitions are induced by the microwave field, the light intensity changes and the transition can be monitored. Such experiments have been performed on $Nd^{3+}:LaCl_3$. Another way to measure hyperfine interactions in excited states is to photoexcite them to a considerable degree and then perform ESR experiments directly [ClHu71, HeHu73].

In recent years a number of techniques employing lasers have been developed that allow the direct investigation of hyperfine interactions by optical means: Optical free induction decay [GeMa76], photon echos [ChHa76, Ta76, TaSz74, Sz70], fluorescence line narrowing [Er75, Er66, PePe77, DePe77], and enhanced and saturated absorption spectroscopy [Er77]. Although these techniques have primarily been used to study the intrinsic linewidth of the optical transitions the hyperfine structure of the $^1D_2 \rightarrow {}^3H_4$ transition in $Pr^{3+}:LaF_3$ [Er77], $Pr^{3+}:LaCl_3$ [PePe77], and Pr^{3+} : $LaBr_3$ [DePe77] has also been investigated; it was also shown that the nuclear spin-lattice relaxation times could be investigated by these techniques.

These developments are too new to have yielded hyperfine interaction information beyond that already known. We can, however, expect that with the development of these techniques and with their application to other crystals, considerable information on the hyperfine interaction in the excited terms of REs will become available.

Here it should be mentioned that resolved hyperfine structure in an undiluted paramagnetic salt ($Ho(C_2H_5SO_4)_3 \cdot 9H_2O$)) was first detected by optical spectroscopy [GrHe61, HeHü62]. This is made possible by a very slow spin relaxation which is a consequence of the large anisotropy in the g-factor of the groundstate crystal field level ($g_1 = 0$). The investigation of resolved hyperfine structure in paramagnetic salts was later the domain of Mössbauer effect experiments [Bl65].

CHAPTER 8

Jahn–Teller Systems

8.1 Introduction

So far it has been assumed that the interaction of the electronic and phonon states does not change the energy in either system (except for a small temperature-dependent contribution, eq. (4.15)). For many crystals this is a valid assumption. In crystals with an orbitally degenerate (not Kramers degenerate) ground state this doubly degenerate state is not the true ground state of the system at $T = 0$. Rather, the interaction of the phonon system and the electron system produces a small distortion which acts to reduce the symmetry of the systems, splitting the orbitally degenerate level. The crystal thereby attains a lower energy, which is then the true ground state of the system. This is the Jahn–Teller effect, which was predicted theoretically in 1937 for free molecules [JaTe37].

The arguments presented by Jahn and Teller also hold for molecules in a crystal. Here the distortion in the position of the central ion and its ligands will interact with that of the neighboring molecules via the elastic properties of the crystal resulting in an interaction between all the ions of the crystal. This interaction is mediated by the local strain, and thus by phonons, as well as by macroscopic strain. Such interaction of the ions in the crystal must produce a phase transition if the temperature is low enough; this phase transition is called a cooperative (because it involves the crystal as a whole and not only certain molecules) Jahn–Teller effect. It can be viewed as an alignment of all the distortions in the crystal, which macroscopically lowers the crystal symmetry and reduces the energy of the electron system. Conceptually the cooperative Jahn–Teller effect is very similar to a magnetic phase transition; instead of the spins that order in the latter case, we now have an ordering of the charge clouds of the 4f ions with respect to a certain crystal orientation. Essential in the magnetic case is the possibility of two spin directions; equally essentially for the Jahn–Teller transition is the possibility of (at least) two orientations of the charge clouds, one of which is gradually stabilized below the phase transition temperature.

These arguments apply only to states that have a degeneracy because of the symmetry in the system. Thus Kramers degenerate levels as such are not influenced by the Jahn–Teller effect, which means that in odd electron systems this effect can manifest itself only in a splitting of a fourfold degenerate level into two twofold degenerate levels. For REs, a level that is degenerate because of symmetry reasons always has an orbital degeneracy because of the large spin–orbit coupling.

The Jahn–Teller transition is a consequence of the electron–phonon interaction, for which the Hamiltonian in lowest order written in spin Hamiltonian formalism is [ElGe71, GeGe75]

$$\mathscr{H}_{\mathrm{JT}} = CQ\bar{S}_z \tag{8.1}$$

where C is the coupling constant between the vibrational states (characterized by the operator Q) and the electronic states (characterized by a pseudo spin operator $\bar{S}_z$). (This equation derives from the first term in eq. (4.6c), given that ε is proportional to the vibrational normal coordinate Q, and that P_{kq} for $k = 1$ is proportional to the first power of the pseudo spin.) The vibrational energy can be written in the harmonic approximation

$$\mathscr{H}_{\mathrm{v}} = \tfrac{1}{2}m\omega^2Q^2 \tag{8.2}$$

Thus with respect to Q, the potential energy of the system has two minima with depth

$$E_{\mathrm{JT}} = C^2/2m\omega^2 \tag{8.3}$$

at $Q = \pm Q_0 = C/m\omega^2$. Above the phase transition, the system tunnels between these two states and one energy level at $Q = 0$ is observed. Below the Jahn–Teller transition the system is stabilized in either of the two minima.

The Jahn–Teller effect is generally quite small in RE compounds, presumably because of the smallness of C. In the RE zircons ($REVO_4$, $REAsO_4$, $REPO_4$), however, the phase transition has often been observed and has been reported on extensively in the literature [GeGe75]. Therefore we shall deal with it here. (Though Jahn–Teller effects have also been observed in other RE systems, the investigation of the zircons seems to be most complete.)

The high temperature symmetry of the RE zircons is tetragonal. Some undergo a structural phase transition at low temperature and the symmetry is then orthorhombic. In addition to the Jahn–Teller transition at T_D, some zircons also show a transition to an antiferromagnetically ordered state at T_N:

$DyVO_4$ with $T_D = 14$ K and $T_N = 3.1$ K [CoMa71]

$DyAsO_4$ with $T_D = 11$ K and $T_N = 2.4$ K [KaKl71]

$TbAsO_4$ with $T_D = 28$ K and $T_N = 1.5$ K [WüBö72, WüLa73, BeKa73a]

$TbVO_4$ with $T_D = 33$ K and $T_N = 0.6$ K [WeWo72, KaSi74].

Two other compounds, $TmAsO_4$ and $TmVO_4$, undergo only a Jahn–Teller transition at $T_D = 6.1$ K [CoMa72] and $T_D = 2.1$ K, [CoSw72] respectively. Finally, some compounds of this class show only magnetic order:

$DyPO_4$ [WrMo71b] with $T_N = 3.4$ K,
$HoPO_4$ [CoSw73] with $T_N = 1.4$ K

$HoAsO_4$ [BeKa73b] with $T_N = 0.56$ K

8.2 Hamiltonian and Molecular Field Approximation

The Hamiltonian describing the Jahn–Teller interaction can be written in the same way, using operator equivalent techniques, as the crystal field Hamiltonian; in more general notation eq. (8.1) reads

$$\mathscr{H}_{JT}(n) = \sum_{kq} C_{kq} Q_{kq}(n) \tilde{O}_{kq}(n) \tag{8.4}$$

where n denotes the unit cell, C_{kq} is the coupling parameters, and $Q_{kq}(n)$ and $\tilde{O}_{kq}(n)$ are the normal coordinates and the electronic tensor operator equivalents, respectively.

To facilitate actual calculation one makes use of the fact that the Jahn–Teller transition occurs only at low temperatures, when only a few electronic energy levels are populated. Then we use the spin Hamiltonian formalism and write

$$\mathscr{H}_{\mathrm{JT}}(n) = CQ(n)\bar{S}_z(n) \tag{8.5}$$

where $\bar{S}_z(n)$ is a pseudo spin operator, which for a doubly degenerate level has eigenvalues $\bar{S}_z = \pm 1$.† In what follows, we shall restrict ourselves to this simple case. In addition, only the molecular field solution will be treated.

Thus there is in the high temperature phase an orbitally degenerate doublet that is split by the Jahn–Teller transition into two singlets at low temperatures; this situation applies to and describes well the situation in $TmVO_4$, with a transition temperature of $T_D = 2.1$ K. There are two contributions to the Jahn–Teller interaction, one due to macroscopic strain and one due to microscopic strain (optical and acoustical phonons). The Hamiltonian due to macroscopic strain is written

$$\mathscr{H}_s = \sum_n C_s \varepsilon \bar{S}_z(n) + \tfrac{1}{2} N\Omega c_s \varepsilon^2 \tag{8.6}$$

where C_s is the elastic strain coupling, c_s the elastic constant corresponding to the strain ε, Ω the volume of the unit cell, and N the number of unit cells. The second term is the macroscopic strain energy. We find the strain by minimizing $\mathscr{H}_s$ with respect to ε:

$$\varepsilon = -\frac{1}{N\Omega c_s} \sum_n C_s \bar{S}_z(n) \tag{8.7}$$

Substituting this back into the original equation yields

$$\begin{aligned} \mathscr{H}_s &= -\frac{1}{2}\left(\frac{1}{N\Omega c_s}\right) C_s^2 \sum_{n,m} \bar{S}_z(n)\bar{S}_z(m) \\ \mathscr{H}_s &= -\frac{1}{2}\frac{\mu}{N} \sum_{n,m} \bar{S}_z(n)\bar{S}_z(m) \end{aligned} \tag{8.8}$$

where $\mu = C_s^2/c_s\Omega$ is the molecular field parameter for the interaction. This Hamiltonian has the same form as the Ising Hamiltonian found in the case of magnetism.

† Note that the choice of the value of the pseudo spin $\bar{S}$ is not in line with the definition given in Section 3.3, which would suggest $\bar{S} = \frac{1}{2}$ for a twofold degenerate state. Since most authors have chosen the given convention, we shall follow it.

Deducing the equivalent Hamiltonian for the coupling to phonons is more involved. Only the final result will be reproduced here; the reader is referred to the papers by Elliott *et al.* [ElGe71, GeGe75] for details.

This Hamiltonian reads (for one ion per unit cell, and an exchange-type interaction)

$$\mathscr{H}_{\mathrm{p}} = \sum_k h\omega(\mathbf{k})(\gamma_k^+ \gamma_k + \tfrac{1}{2}) - \tfrac{1}{2} \sum_{n,m} J_{\mathrm{p}}(n-m)\bar{S}_z(n)\bar{S}_z(m) \tag{8.9}$$

where $\omega(\mathbf{k})$ is the phonon frequency with wavenumber $\mathbf{k}$, γ_k and γ_k^+ are operators that have been obtained by a linear transformation of the phonon creation and annihilation operators, and J_{p} gives the coupling, from which the self-energy has been subtracted. Thus, a second term is generated, which again has the form known from the theory of magnetism. The total Jahn–Teller Hamiltonian now yields

$$\mathscr{H}_{\mathrm{JT}} = \sum_k h\omega(\mathbf{k})(\gamma_k^+ \gamma_k + \tfrac{1}{2}) - \tfrac{1}{2} \sum_{n,m} \{J_{\mathrm{p}}(n-m) + (\mu/N)\}\bar{S}_z(n)\bar{S}_z(m) \tag{8.10}$$

From the molecular field approximation we obtain

$$\mathscr{H}_{\mathrm{JT}} = -(\lambda + \mu)\langle \bar{S}_z \rangle \sum_n \bar{S}_z(n) \tag{8.11}$$

where $\mu = C_{\mathrm{s}}^2/c_{\mathrm{s}}\Omega$ and $\lambda = (1/N)\sum_{n \neq m} J_{\mathrm{p}}(n-m)$. This Hamiltonian has the energy values (for $\bar{S} = 1$)

$$W = \pm(\lambda + \mu)\langle \bar{S}_z \rangle \tag{8.12}$$

With Boltzmann statistics this yields the following relations

$$\begin{aligned}
&\text{Order parameter:} && \langle \bar{S}_z \rangle = \tanh(W/k_{\mathrm{B}}T) \\
&\text{Splitting of the states:} && \Delta = 2W = 2(\lambda + \mu)\tanh(W/k_{\mathrm{B}}T) \\
&\text{Transition temperature:} && T_{\mathrm{D}} = (\lambda + \mu)/k_{\mathrm{B}}
\end{aligned} \tag{8.13}$$

8.3 Typical System: $TmVO_4$

Of the many Jahn–Teller systems investigated so far, $TmVO_4$ is perhaps the simplest; therefore it will be used to describe the behavior of an actual system. At high temperature the symmetry of $TmVO_4$ is tetragonal (D_{4h}^{19}); at $T_{\mathrm{D}} = (2.15 \pm 0.15)$ K the substance undergoes a Jahn–Teller transition and the symmetry is reduced to orthorhombic (D_{2h}^{24}). The site symmetry for the Tm^{3+} ion changes from D_{2d} to D_2. Spectroscopic measurements show that for $T > T_{\mathrm{D}}$ the lowest crystal field is a doublet and the next higher level is a singlet at 50 cm^{-1}. Therefore, this system is a very good model for a Jahn–Teller system with a pseudo spin $\bar{S}_z = \pm 1$. Below T_D, due to the lowering of the site symmetry, the doublet is split.

The realization that some RE zircons show a Jahn–Teller transition originally came from specific heat measurements [CoEl70]. But the first direct determination of the RE site symmetry in the Jahn–Teller state was done by spectroscopic measurements, which are suited for this kind of investigation because they show a change of selection rules when the temperature is lowered from above to below T_D [BeLa71].

The temperature dependence of the splitting of the doubly degenerate ground state level in $TmVO_4$ could be measured by optical spectroscopy [BeLe72b]. Figure 38 shows the result of that experiment by Becker *et al.* [BeLe72b]; the optical data compare very well with the magnetic data [CoSw72], which are shown for comparison. The accuracy of the spectroscopic data above T_D is limited by the linewidth, which makes the relatively large error bars in that region understandable. The straight line is a fit to the molecular field equation (Δ is the total splitting of the Jahn–Teller split doubly degenerate ground level; $\Delta = 2W$)

$$\Delta = \Delta_0 \tanh W/k_B T \tag{8.14}$$

yielding $T_D = (2.14 \pm 0.04)$ K, and $2W_0 = \Delta_0 = (2.97 \pm 0.04)$ cm^{-1}, whence $k_B T_D / W_0 = 1.00 \pm 0.04$. The good fit of the experimental data

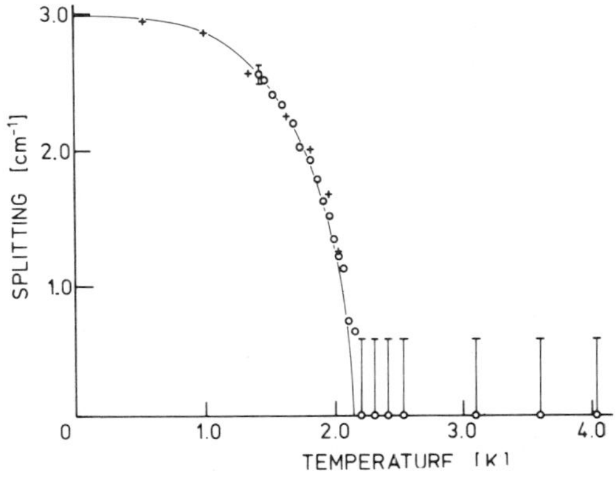

Fig. 38. Temperature dependence of the ground state splitting in $TmVO_4$ deduced from the $I^3H_6 \rightarrow a^1G_4$ transition.

The circles are spectroscopic data, the crosses are magnetic data. The line is a fit of eq. (8.14) to the experimental data points with $T_D = 2.14$ K and $\Delta_0 = 2.97$ cm^{-1} [BeLe72b] (Copyright by The Institute of Physics)

to the molecular field equation and the fact that the ratio $k_B T_D/W_0$ is unity suggest that the interactions in this system are very long range (because only then would the molecular field approach give good results), which in turn means that the Jahn–Teller transition is induced predominantly by macroscopic strain. This is confirmed by other experiments (GeGe75).

A similar conclusion can be reached by performing an experiment under external applied stress σ [BaKa75a]. In the phase above T_D, the lobes of the charge distribution can be thought of as pointing in the [100] direction. Below the transition temperature T_D they can point in either the [110] or the [1$\bar{1}$0] direction. With zero applied stress, both directions are equally likely, and thus both possibilities are present as domains. The two possible directions mean that the order parameter $\langle \bar{S}_z \rangle$ is either positive or negative. One can, however, stabilize one of the distortions and enhance it by applying stress σ in the [110] or [1$\bar{1}$0] direction. This experiment was performed by Battison *et al.* [BaKa75a] with the result given in Fig. 39, which shows the splitting of the ground state as a function of [110] stress at a temperature close to T_D. To incorporate the externally applied stress as described in the preceding section, add it to the molecular field Hamiltonian (8.11) to obtain

$$\mathscr{H}_{JT} = -(\lambda + \mu)\langle \bar{S}_z \rangle \sum_n \bar{S}_z(n) - (\mu\Omega/c_s)^{1/2}\sigma \sum_n \bar{S}_z(n) \qquad (8.15)$$

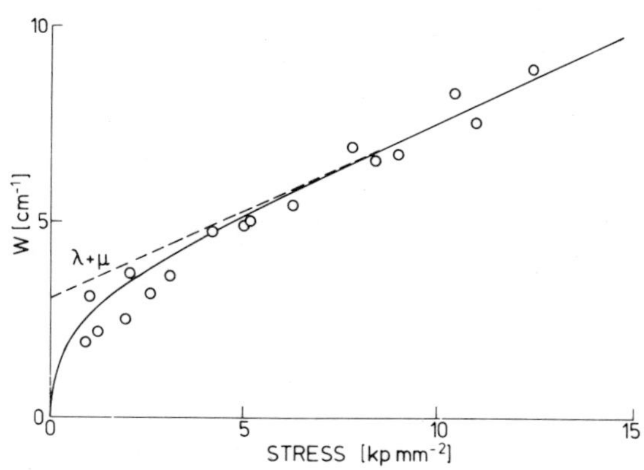

Fig. 39. Variation of the ground state splitting in $TmVO_4$ for $T = T_D$ as a function of stress along the [110] direction; the circles are from optical absorption data.

The straight dashed line should intercept the axis at the zero temperature splitting $\mu + \lambda = 3\ cm^{-1}$. The solid curve is the molecular field result from eq. (8.16) [BaKa75a] (Copyright by The Institute of Physics) where the constants are defined as above. The eigenvalues are now

$$W = (\lambda + \mu)[\langle \bar{S}_z \rangle + (\sigma/\sigma_0)]; \qquad \sigma_0 = (\lambda + \mu)(c_s/\Omega\mu)^{1/2}$$
$$\langle \bar{S}_z \rangle = \tanh[W/k_B T] \tag{8.16}$$

The solid curve in Fig. 39 was calculated by using these equations with parameters ($\lambda + \mu = \Delta_0/2 = 1.5\ \text{cm}^{-1}$, $c_s = 1.68 \times 10^{11}$ dyn cm^{-2} obtained from ultrasonic measurements [MePy73]). The best fit to the data was obtained by setting $\lambda/\mu = \frac{1}{3}$. This means that the coupling to the optical phonons is zero [HaHa72], and therefore only the coupling to the macroscopic strain occurs, making the good description of the experimental data by the simple molecular field treatment understandable.

The twofold degenerate ground state in $TmVO_4$ (above T_D) also has a magnetic moment. Thus at low temperatures the Jahn–Teller and the magnetic interactions compete: In $TmVO_4$ the former wins and the system undergoes a distortive transition. The distortion is along the z axis (eq. (8.16)); the g-tensor is uniaxial and is directed along the x axis. Therefore, we can write the total Hamiltonian with the Jahn–Teller and the Zeeman energy

$$\mathscr{H} = \tfrac{1}{2} g_x \mu_B H_x \sum_n \bar{S}_x(n) - (\lambda + \mu)\langle \bar{S}_z \rangle \sum_n \bar{S}_z(n) \tag{8.17}$$

(The factor $\frac{1}{2}$ in the first term enters because we describe the two level system by $\bar{S} = 1$ instead of $\bar{S} = \frac{1}{2}$.) This Hamiltonian shows that the two interactions are quantized with respect to orthogonal axes. For $T > T_D$ the order parameter $\langle \bar{S}_z \rangle$ is zero and for $H \| x$ the normal Zeeman effect is observed. For $T < T_D$ and $H > 0$, the Zeeman energy and the Jahn–Teller energy compete with each other. Then the direction of the pseudo spin lies in the x–z plane at an angle α with respect to the z axis, with $\alpha = 0$ for $H = 0$ and $\alpha = \pi/2$ for $H > H_c$ (a critical value that we will calculate next). Thus the magnetic field pulls the pseudo spin from the z axis to the x axis, and if it is parallel to the x axis, the distortion transition has been suppressed by the magnetic field.

The Hamiltonian (8.17) enables us to calculate the energy

$$W = \pm[(\tfrac{1}{2} g_x \mu_B H_x)^2 + ((\lambda + \mu)\langle \bar{S}_z \rangle)^2]^{1/2} \tag{8.18}$$

This equation is similar, although not equal, to eq. (3.47b) because the crystal field term contains $\langle \bar{S}_z \rangle$, the order parameter, which is also a function of the applied magnetic field.

Now according to Gehring *et al.* [GeMo72] we have

$$\langle \bar{S}_z \rangle = \langle \bar{S}_\alpha \rangle \cos\alpha \tag{8.19}$$

where

$$\cos\alpha = \frac{[W^2 - (\frac{1}{2}g_x\mu_B H)^2]^{1/2}}{W}$$

and

$$\langle \bar{S}_\alpha \rangle = \tanh W/kT$$

Inserting this back into (8.18), we find

$$W = \pm(\lambda + \mu)\tanh(W/k_BT) \qquad \text{for} \quad \tfrac{1}{2}g_x\mu_B H_x < W(H_x = 0)$$

$$W = \pm\tfrac{1}{2}g_x\mu_B H \qquad \text{for} \quad \tfrac{1}{2}g_x\mu_B H_x > W(H_x = 0)$$

This means that for $\frac{1}{2}g_x\mu_B H_x < W(H_x = 0)$ the splitting of the ground doublet is field independent; an exact compensation of Jahn–Teller and

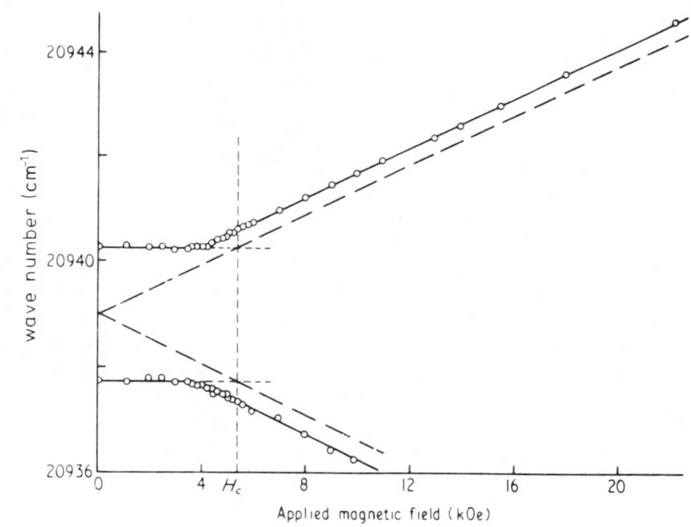

Fig. 40. Ground state splitting in $TmVO_4$ at 1.4 K as a function of magnetic field applied Parallel to the [001] axis as measured by optical absorption in the 20 939 cm^{-1} line.
The heavy broken line has been corrected for demagnetization and dipole fields and is extrapolated to zero external field; the critical field H_c is indicated by the thin broken line [BeLe72b] (Copyright by The Institute of Physics)

Zeeman energies takes place; at the critical field

$$H_c = \frac{2(\lambda + \mu)}{g_x \mu_B} \tanh(g_x \mu_B H_x / 2k_B T)$$

the pseudo spin is aligned with the x axis, $\langle \bar{S}_z \rangle = 0$, and for higher fields only the normal Zeeman effect is observed. These considerations are borne out by the experimental results [BeLe72b] shown in Fig. 40, where the behavior of a typical absorption line in $TmVO_4$ below T_D under the application of an external magnetic field parallel to the x axis is shown. The solid lines were calculated taking demagnetization effects into account, whereas for zero demagnetization the broken lines were obtained. The agreement between theory and experiment is good.

The change in symmetry at T_D from tetragonal to orthorhombic changes the crystal from an optically uniaxial one to a biaxial one. This can be viewed directly by using polarized light [BeLe72b]. Indeed, this symmetry change enables us to determine the axis of the distortion directly. The change of optical isotropy can also be used to measure the order parameter $\langle \bar{S}_z \rangle$ by linear birefringence [BeGe75, Be76]. Above T_D the optical axis is the tetragonal axis and the index of refraction is isotropic in the plane perpendicular to it. Below T_D the index of refraction is anisotropic in this plane and it can be shown that $\delta n = n_x - n_y$ is proportional to $\langle \bar{S}_z \rangle$, which can be accurately measured by measuring the linear birefringence.

Despite the brevity of this section, we hope we have transmitted the flavor of this very interesting phenomenon.

CHAPTER 9

Divalent Rare Earth Compounds: The Europium Chalcogenides

9.1 Introduction

So far our discussion has been restricted to the trivalent REs in insulators. Some REs, however, notably Sm, Eu, and Yb, also occur as divalent ions in crystals, and we shall discuss some of the basic properties of their spectra here (for a review of spectra of divalent REs see, e.g., [McClKi63, KiYo64]). The intra-4f transitions in compounds containing divalent REs are very similar to those of the corresponding trivalent REs (where the former have an additional 4f electron, of course). The main difference is a small compression of the energy scale brought about by smaller Coulomb and spin–orbit interactions, which are caused by the smaller effective nuclear charge of the divalent ions. Because of this situation the intra-4f spectra of divalent RE ions need not be dealt with separately here.

The visible absorption spectra of divalent REs are noticeably different from those of the trivalent REs. Figure 6 shows that the $4f^N \rightarrow 4f^{N-1}\,5d$ transition energies, which for RE^{3+} ions are in the ultraviolet spectral region, are well within the visible region for RE^{2+} ions. These transitions are not parity forbidden, and are therefore about 10^6 times stronger than the $4f \rightarrow 4f$ transitions. In the region in which they occur, they can completely swamp the weak 4f → 4f transitions, and for divalent REs we are limited to observing them mostly in the red and infrared spectral regions. The visible spectra of a divalent RE compound like EuF_2 are dominated by the $4f^7 \to 4f^6 5d$ transitions.

One class of divalent RE compounds, the Eu chalcogenides (EuO, EuS, EuSe, EuTe) has been investigated very thoroughly [Wa73, Me65, Wa71, Wa77] because of their simple NaCl structure and because they are semiconductors that order magnetically (magnetic semiconductors) [MaBo61, BuJu62, McGuAr62, Ho62, MeMa68]. Though no longer transparent in the rigorous sense of the word, they can be used for absorption measurements as thin films. Also, their properties have been so extensively investigated that they can serve to demonstrate the spectroscopic properties of substances, in which the spectra are dominated by the 5d electron.

Originally it was hoped that by doping, the ordering temperature of these compounds, which is below 100 K, could be raised to above 300 K. This would have made them interesting for technical applications. This hope has been abandoned, but even without new technological applications these compounds have enhanced our understanding of ferromagnetism in nonmetals, of which they are the best documented example.

9.2 Crystal Field Interaction of One 5d Electron

In principle, the crystal field theory outlined in Chapter 3 can also be used to describe the crystal field interaction of a 5d electron. For historic reasons, the notation for d electrons differs from that for 4f electrons; we shall mention a few relevant facts [Gr64, AbBl70]. The 5d electrons extend spatially farther out than the 4f electrons, which means that the crystal field interaction for a 5d electron is larger than for a 4f electron; in fact, in comparison to it the spin–orbit interaction can be neglected in first order.

For an octahedral (cubic) coordination, the crystal field Hamiltonian in spin Hamiltonian notation has the form (we keep only the $k = 4$ terms; the $k = 2$ terms are zero for reasons of symmetry and the $k > 4$ terms are zero because $k \leq 2l$; see Section 3.2) [Hu64]

$$\mathscr{H}_{CF} = A_{40}\langle r^4\rangle\beta(O_{40} + 5O_{44}) \tag{9.1}$$

where one often writes

$$A_{40}\langle r^4\rangle = Dq/12; \qquad \beta = 2/63 \tag{9.2}$$

Since due to its smallness the spin–orbit interaction can be neglected in first order, the states of a 5d electron in a cubic crystal field are calculated by applying the Hamiltonian (9.1) to its fivefold orbitally degenerate state (^{2}D). This means the basic functions are those characterized by L and M_L instead of J and M_J, as in the case of the 4f electrons. The matrix elements of (9.1) can be calculated with the help of Table 8b, and solving the resulting secular equation yields energies at $-4Dq = (-96/63)A_{40}\langle r^4 \rangle$, which is an orbital triplet state, and at $+6Dq = (144/63)A_{40}\langle r^4 \rangle$, which is an orbital doublet state. The total splitting between the triplet and doublet state is $10Dq = (80/21)A_{40}\langle r^4 \rangle$. The triplet state is a Γ_5 state and the doublet state a Γ_3 state. More conventional, however, is the molecular orbital terminology in which the Γ_5 state is called a t_{2g} state and the Γ_3 state an e_g state.

For configurations of more than one 5d electron, the splitting pattern gets more complicated, but we shall not need such configurations in our discussion (see [SuTa70]).

9.3 Absorption Spectra of the Europium Chalcogenides

The ground term of the Eu^{2+} ion is $4f^7\ ^8S_{7/2}$, the same as that of Gd^{3+}. The excited states accessible by optical spectroscopy are either $4f^7$ or $4f^6 5d$, both of which fall in the visible region. Inspecting the absorption spectra of Eu^{2+}:SrS, Eu^{2+}:KBr, and EuF_2 [ReGl64, FrHo68, FrMe68] in Fig. 41, we see that they are dominated by two broad peaks, very different from the typical RE^{3+} ion spectra seen, e.g., in Fig. 1. It is therefore appealing to ascribe the spectra in Fig. 41 to $4f^7 \to 4f^6 5d$ transitions and to assume that on the coarse scale of the figure the $4f^7 \to 4f^7$ transitions do not show up at all. The interpretation of a $4f^N \to 4f^{N-1} 5d$ transition is not quite so easy as that of an intra-4f transition; therefore we shall deal with it in more detail. For a first order interpretation (a better description is given later), an energy level diagram of the $4f^6 5d$ configuration is constructed by treating the crystal field and intra-atomic interactions of the $4f^6$ and 5d configurations separately, and then merely summing the energies of the two electronic systems. The 5d electron is split into a t_{2g} state and an e_g state in a cubic environment, which is present in SrS and EuF_2 and can be taken as a first approximation in KBr (the charge compensation center destroys the cubic symmetry at the Eu^{2+} site in KBr); the order of the t_{2g} and e_g states in EuF_2 is the opposite of that in SrS.

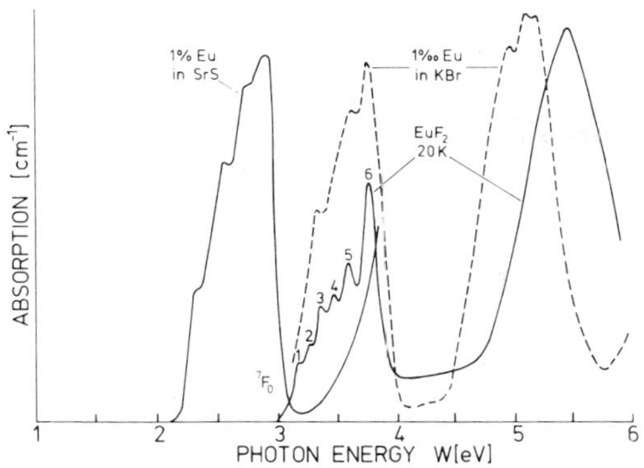

Fig. 41. Absorption spectra of Eu^{2+} : SrS, Eu^{2+} : KBr, and EuF_2 [FrMe68, Wa73]. Each spectrum shows the dominant two peak structure, which is produced by the $e_g - t_{2g}$ splitting of the 5d electron. (The second peak is not visible for Eu^{2+} : SrS due to the onset of the intrinsic absorption in SrS). The lower peak in the EuF_2 spectrum shows the superimposed structure of the $f^6\ {}^7F_J$ ground state.

The crystal field interaction in the $4f^6$ configuration is small compared to that in the 5d state, and the energy diagram is in first order that of the free ion for $4f^6$, the ${}^6F_J$ multiplet. The lowest $4f^6 5d$ energies should therefore consist of a t_{2g} and an e_g state and on each of them the ${}^7F_J$ multiplet structure is superimposed. This is indeed clearly seen in the lower absorption peak of the EuF_2 spectrum, substantiating the foregoing discussion. In the other peak, the $4f^6$ structure cannot be so nicely resolved. There may be various reasons for this difficulty such as the spin–orbit coupling in the 5d state, deviations from the cubic symmetry (by the charge compensating center in KBr), or lifetime effects. The absorption spectrum is, however, dominated by the strong absorption provided by the 5d electron and its crystal field interaction. The total crystal field interaction of the 5d electron as measured by the separation of the two peaks is about 1.5 eV or 12,000 cm^{-1}; by comparison with the splitting of a $J = \frac{3}{2}$ term in Er^{3+} compounds, which is of the order of 100 cm^{-1}, the crystal field interaction of a 4f electron differs from that of a 5d electron by two orders of magnitude. In SrS the second 5d transition is not visible because of the onset of the strong intrinsic absorption in SrS.

Inspecting the absorption spectra of the Eu chalcogenides in Fig. 42 [GüIm71, Gü74], we see that they are similar in their gross structure to those in Fig. 41; that is, they show two strong absorption peaks in the region below 6 eV, which are then interpreted as transitions between the $4f^7$ ground state of Eu^{2+} and the $4f^6 5dt_{2g}$ and the $4f^6 5de_g$ excited states.

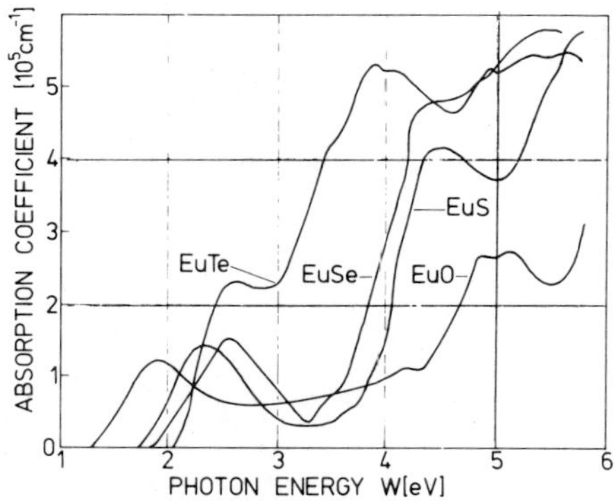

Fig. 42. Absorption spectra of thin films of the Eu chalcogenides [GüIm71]. The main structures are two peaks.

The increasing crystal field splitting in going from EuTe (1.5 eV) through EuSe (1.7 eV) and EuS (2.2 eV) to EuO (3.1 eV) may be rationalized on simple lattice constants arguments. Dq $\sim 1/a_0^5$ and the lattice constant a_0 decreases from EuTe to EuO (experimentally, however, a power law of only -3.5 instead of -5 is found [Wa77]); the center of gravity of the absorption to the two 5d levels stays roughly constant through the series. An energy level diagram for the Eu chalcogenides [GüIm71, EaHo69, FeSc69, DiHa70, Wa71, Wa77] can be constructed (Fig. 43) by starting from the ionic energy levels obtained by applying the appropriate Madelung correction to the free ion energy levels. The details are then filled in so as to give the best agreement with all the available experimental data. The relative positions of the $5f^7$ and the anion p^6 levels can be adjusted from photoemission data. The splitting of the 5d levels is taken from optical absorption experiments and their width is taken such that the observed spectrum is reproduced correctly with the known $4f^6\ {}^7F_J$ structure, leading to 5d widths of 1.1 eV (EuO), 0.7 eV (EuS and EuSe), and 0.6 eV (EuTe). The width of the p^6 anion level is taken from photoemission spectra. For each compound, the zero of the energy level diagram has been placed at the $4f^7$ ground state.

More careful consideration reveals that things are slightly more complicated. The 5d electron interacts strongly with the crystal. As a consequence, it is no longer appropriate to describe this electron in a localized picture, as for 4f electrons, but a band description must be used instead. The 5d states are therefore one-electron band states (in contrast to the 4f states, which are many-electron localized states). A $4f^7 \rightarrow 4f^6 5d$ transition therefore puts an electron into the 5d band and leaves behind

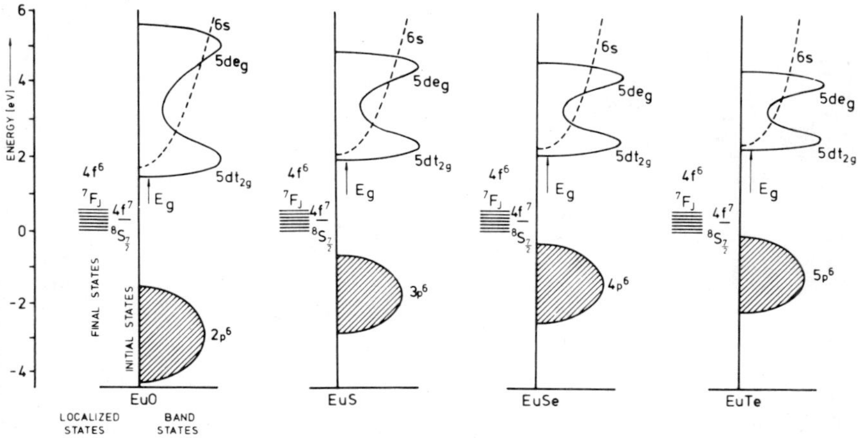

Fig. 43. Energy level diagrams for the Eu chalcogenides [Wa77]. They have been brought into coincidence for the different chalcogenides by lining up the position of the $4f^7$ configuration. The position of the 6s bands has been estimated and is shown by a dashed line.

a localized $4f^6$ state, which may be in either of the ^7F_J ground terms (the next excited configurations are too high to be excited). In order to make clear the distinction between band states and localized states, it has become customary to separate them in an energy level diagram. Thus in Fig. 43 the diagrams at the left show the initial ($4f^7\ ^8S$) and final ($4f^6\ ^7F_J$) localized states for each compound; at the right the p^6, 5d, and 6s band states are shown. There is, however, some ambiguity in this sort of diagram because it does not disclose the relative positions of the f^7 and the f^6 states. Thus the figure could be drawn with the localized and band states in a single diagram. The main difference then would be a superposition of the ^7F_J states upon either one of the 5d states. We have followed the procedure of Wachter [Wa77, Wa71] in differentiating between the two sorts of states.

Above 1 K, all the Eu chalcogenides show spontaneous magnetic order, and Table 41 [Wa77] summarizes their magnetic properties. This table shows that the strength of the magnetic interaction has the same systematics as the strength of the crystal field interaction: The size of both these interactions is proportional to the radial extension of the 5d electrons. On further analysis we find that the magnetic interaction can be described by two constants, which describe the interaction with the nearest (nn) and the next nearest (nnn) neighbors. We see that J_{nnn} is positive (antiferromagnetic) and of about constant magnitude, whereas

TABLE 41

MAGNETIC PROPERTIES OF THE Eu CHALCOGENIDES[a]

	a_0 (Å)	Magnetic order	$T_{C,N}$ (K)	$W(e_g - t_{2g})$ (eV)	ΔW_R^b (eV)	J_{nn} (K)	J_{nnn} (K)	Absorption edge (eV)
EuO	5.141	Ferro	69.3	3.1	−0.27	−1.212	−0.238	1.12
EuS	5.968	Ferro	16.6	2.2	−0.18	−0.472	+0.236	1.65
EuSe	6.195	Antiferro	4.6	1.7	−0.07	−0.146	+0.022	1.78
EuTe	6.598	Antiferro	9.4	1.5	+0.03	−0.086	+0.30	2.00

[a] From [Wa77, PaDi76]. Note that we use $J_{12}\mathbf{S}_1 \cdot \mathbf{S}_2$ instead of $-2J_{12}\mathbf{S}_1 \cdot \mathbf{S}_2$ as Hamiltonian for the exchange interactions.

[b] ΔW_R is the red shift of the absorption edge.

J_{nn} is negative (ferromagnetic) and varies strongly: The latter integral is thought to come about by indirect cation–cation superexchange via the 5d orbitals, and is therefore strongly dependent on the radial extension of these orbitals.

Of particular interest in the Eu chalcogenides is the position of the absorption edge (called the gap energy) as a function of temperature. When the temperature is lowered, the absorption edge of a semiconductor usually exhibits a small blue shift, a consequence of the decreasing electron–phonon interaction. In the paramagnetic region the Eu chalcogenides do indeed show this expected behavior. If the temperature approaches the Curie (or Néel) temperature, a red shift of the absorption edge starts to appear that gets strong below the ordering temperature. This red shift is quite large—0.27 eV in EuO [BuWa66, BuJu64], which is about 20% of a room temperature gap energy of 1.2 eV. The temperature dependence of the gap energy is given in Fig. 44. This figure shows that the shift of the absorption edge does not follow the magnetization curve, which would make the position constant above the ordering temperature. For EuO and the other chalcogenides (as for the absorption line positions in DAG; see Chapter 6) the shift follows the size of the short range order parameter (spin correlation function), which is given by the normalized magnetic part of the specific heat (see Fig. 44). The red shift can be increased by a magnetic field. Since in a ferromagnet like EuO an external field increases the magnetization of the sample, the red shift is a function of the spin correlation function. The shift is -0.27 eV at saturation, corresponding to $\sim 2000\ \mathrm{cm}^{-1}$; this is clearly too much to be caused by a magnetic splitting in the ground state, which is of the order of the ordering temperature, $70\ \mathrm{cm}^{-1}$.

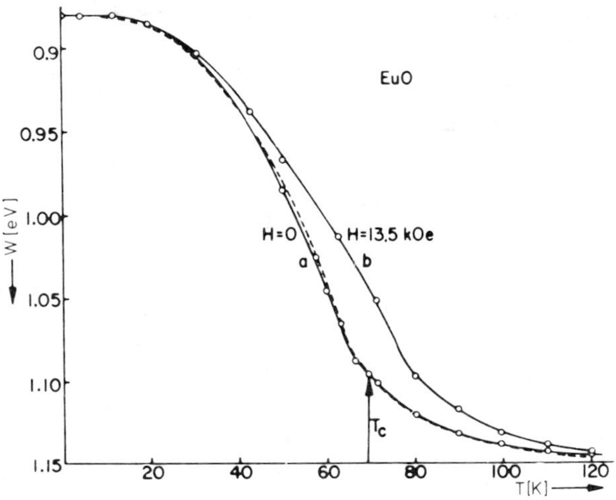

Fig. 44. Temperature dependence of the energy gap in EuO.
The data points give the energy of the gap. The solid curve connects the experimental optical data points. The dashed curve gives the temperature dependence of the magnetic specific heat, normalized in such a way such as to agree at high and low temperatures with the gap energy. As in DAG (Fig. 36), good agreement between the temperature dependence of the magnetic specific heat and the gap energy are seen [Wa73]. With an applied magnetic field the magnetic energy and the short range order increase (EuO is a ferromagnet). This holds also for the red shift of the energy gap.

The basic features of the magneto-optical spectra of the Eu chalcogenides seem to be understood [Wa77, Ka73]. The largest interaction experienced by the 5d electron is the crystal field interaction, which gives rise to a roughly temperature independent splitting, with the t_{2g} level the lowest one. The exchange interaction that produces the observed red shift of the absorption edge is between the spin ($S = \frac{1}{2}$) of the t_{2g} electron and the spin ($S = 3$) fo the ^7F_J final state; it yields two states, with spin $S = \frac{7}{2}$ and $S = \frac{5}{2}$, respectively, split by twice the red shift ($2\Delta W_R$). Finally, for a complete energy level diagram, the spin–orbit interaction of ~ 0.25 eV has to be taken into account.

Kasuya [Ka73] uses a slightly different nomenclature. If it is acknowledged that the 5d electron is neither a localized electron nor a real band electron (because it will interact with the 4f hole left behind by the excitation), the 5d state can be called an exciton state. Since it is an exciton in magnetic ordered compounds, Kasuya calls the state a magnetic exciton.

The optical spectra of some divalent Eu compounds (Eu^{2+}:KCl, Eu^{2+}:NaCl, EuO, EuS, EuSe, EuTe) have also been investigated by using the Faraday effect and magnetic circular dichroism [Fe74, Schoe75]. The measurements have essentially confirmed the conclusions obtained by the combination of absorption spectroscopy and photoemission data, yielding a spin–orbit splitting of ~0.25 eV in the t_{2g} state, in agreement with the result suggested earlier.

9.4 A Mixed Valence Compound: Eu_3O_4

In most compounds REs are trivalent and in some they are divalent. Yet in some substances called mixed valence compounds, both divalent and trivalent RE ions are present (for a review of this topic see [Va76, WoCo73]). The most common mixed valence RE compounds are metallic and do not fall within the scope of this book. Some, however, are semiconductors (or even insulators), and optical investigation of them has produced much insight into the energy level diagrams of these substances; therefore an example of these compounds is discussed here.

The class of mixed valence compounds can be divided into three groups [BaKa76b]. In the two most common forms the RE ions occupy equivalent lattice sites. In these cases a change in valence can occur either by a temperature activated hopping process (Sm_3S_4, Eu_3S_4) or, if the sample is metallic, by the exchange of an electron between the 4f shell and the conduction band (SmB_6, metallic SmS). In a third group of compounds the RE occupy inequivalent sites, so that no charge exchange can occur between the divalent and trivalent ions (Eu_3O_4). The 4f electrons are quite localized even in a metal, and more so in the semiconducting state. Therefore, if we know the energy level diagram for the RE ions in a mixed valence compound (even one which has no charge exchange) like Eu_3O_4 [BaKa75b], this diagram can be used as a reference for the semiconducting and metallic compounds.

In Eu_3O_4 ($Eu^{2+}Eu_2^{3+}O_4^{2-}$) the divalent and trivalent Eu ions have a ratio of 1:2. The compound has the $CaFe_2O_4$ [Ra66] structure, the Eu^{2+} ions occupying the Ca sites and the Eu^{3+} ions occupying the Fe sites. Electrical measurements show that Eu_3O_4 is a reasonably good insulator [BaKa75b]. Since Eu_3O_4 cannot be grown in thin films, its opaqueness in the visible and ultraviolet regions prevents our taking other than reflection spectra. From them via a Kramers–Kronig analysis the imaginary part of the dielectric constant (the absorption) is obtained; the results are shown in Fig. 45, which also gives the analysis of the spectral structure.

Since we again observe very strong absorption, the transition must be parity allowed, meaning that for the 4f electrons we see $4f^N \rightarrow 4f^{N-1}5d$

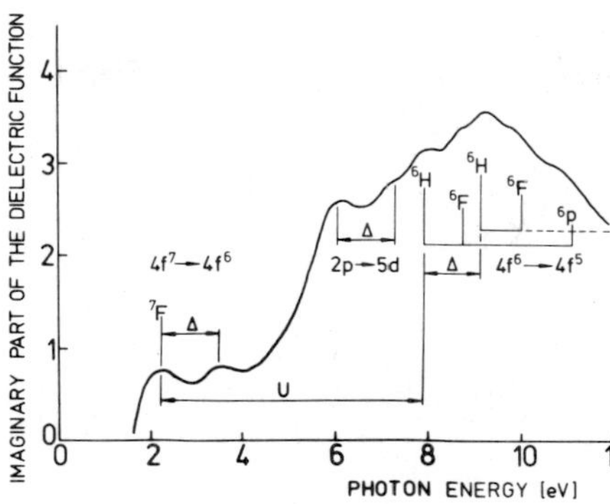

Fig. 45. Imaginary part of the dielectric function for Eu_3O_4.
The crystal field splitting of the 5d band (Δ) and the Coulomb correlation energy (U) are indicated along with the center of gravity of the ^6H, ^6F, and ^6P multiplets with the intensity calculated from coefficients of fractional parentage [Co75b, BaKa76a].

transitions, specifically, $4f^7 \rightarrow 4f^6 5d$ (for divalent Eu) and $4f^6 \rightarrow 4f^5 5d$ (for trivalent Eu). As in the cubic case, the 5d state is split into two levels. In Eu_3O_4 neither the Eu^{2+} nor the Eu^{3+} ions occupy cubic sites, but the crystal field can be approximated by a cubic one, such that the two hump structure is preserved. In a $4f^N \rightarrow 4f^{N-1}5d$ transition the $4f^{N-1}$ structure of the RE ions after the transition (final state) must not be in the ground state, but can also be in some excited state. For the $4f^7 \rightarrow 4f^6 5d$ transition the $4f^6 5d$ final state is $4f^6(^7F_J)5d$ (the next terms are too high energetically to be excited) and for the $4f^6 \rightarrow 4f^5 5d$ transition the $4f^5 5d$ final state is $4f^5$ $(^6H_J, {}^6F_J, {}^6P_J)5d$. As was seen in the discussion of the chalcogenides, the internal ^7F_J structure cannot be resolved in the spectra. One can, however, see the $4f^5$ $(^6H, {}^6F, {}^6P)5d$ final state structure in Fig. 45.

A property uniquely observed in the spectra of mixed valence compounds is the so-called Coulomb correlation energy U. This is the energy difference between the $4f^7$ and the $4f^6$ configuration in solids (generally, the $4f^N$ and the $4f^{N-1}$ configuration); in other words, the energy necessary to produce the $4f^7$ configuration out of the $4f^6$ configuration by putting an extra electron into its 4f shell. In Eu_3O_4 this energy difference is measured between the position of the 4f ground state configurations for Eu^{2+} ions ($4f^7$) and Eu^{3+} ions ($4f^6$) and it comes out to be 5.7 eV. This value is much smaller than we would estimate from free ion spectra, because in the crystal there is screening by the charge of the ligands. Figure 46 shows the energy level diagram consistent with the interpretation of the spectrum in Fig. 45. As for the energy level diagrams of the Eu chalcogenides, the band states (which are one electron states) and 4f states (which

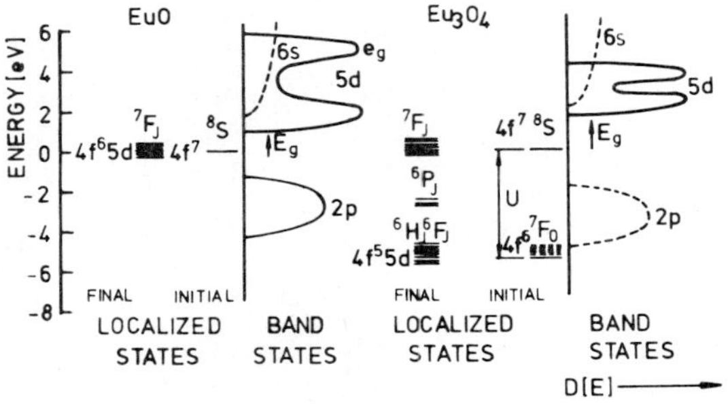

Fig. 46. Energy level diagrams for EuO and Eu_3O_4.
At the right in each diagram are the band states; at left the localized 4f states. For the latter, the initial and final states are distinguished. The diagram for Eu_3O_4 contains the $4f^7$ (Eu^{2+}) and the $4f^6$ (Eu^{2+}) energy levels [BaKa75b]. The position and width of the 6s bands and the 2p band in Eu_3O_4 are only estimated and are therefore shown by a dashed line; the ^7F_J ($J = 1 - 6$) terms in Eu_3O_4 are not occupied in the initial state and therefore given by dashed lines. The label $4f^65d$ has been omitted for the ^7F_J terms of Eu_3O_4.

are many electron states) are separated. At the left, again, the energy level diagram of EuO is shown (see Fig. 43); the corresponding diagram for Eu_3O_4 is shown at the right. It shows the same features for the band states as the EuO diagram, although the 5d crystal field splitting is slightly smaller; in addition to the $4f^7$ ($^8S_{7/2}$) state with its $4f^6$ (^7F_J)5d final state, we can now see the $4f^6$ (^7F_J) ground state with its $4f^5$ (^6H_J, ^6F_J, ^6P_J)5d final state as well. It is separated by the Coulomb correlation energy U from the former state.

Due to the localized nature of the 4f electrons, it is valid to assume that similar energy band schemes will also apply for other RE intermediate valence compounds. This has been verified by an investigation [BaKa76a, BaKa76b] of the energy level scheme of Sm_3S_4 and of SmS (in the semiconducting phase which is purely divalent, and the metallic phase, which is an intermediate valence system). The Sm chalcogenides differ from the Eu chalcogenides in that the $4f^N - 4f^{N-1}5d$ gap is much smaller in them. Thus by applying moderate pressure (6.5 kbar) to SmS, the 5d band can be made to intersect the 4f state [JaNa70]. Then the $4f^6$ state can emit an electron into the 5d band, thereby becoming a $4f^5$ state. Thus the highest occupied level in metallic SmS is one in which the $4f^6$ configuration emits and absorbs an electron, thus changing to a $4f^55d$ configuration. Therefore, we have for metallic SmS, in contrast to the Eu_3O_4, if the time average is taken, a state with a nonintegral valence on every ion.

CHAPTER 10

Rare Earths in Glasses

The preceding chapters have dealt exclusively with crystalline solids. However, many solids that are of practical use today are amorphous or glassy rather than crystalline. This means that the immediate environment of like ions in such substances is similar, but that there is no long range order in the sample. In terms of scattering theory, the two site correlation function is damped out very rapidly, so that it is practically zero beyond a nearest neighbor distance. RE ions can be easily incorporated into many glasses; so these systems represent in principle an interesting area of research. It was noticed quite early that in glasses—as might be expected—the most prominent feature of the RE crystal spectra, the extreme sharpness of the optical lines, vanishes. Therefore, RE ions in glasses have not been studies much for a long time. When it was realized that REs in glasses could be used as laser materials, however, a number of investigations into their spectra and fluorescence properties were undertaken [Re73, Re75]. In general, the spectral lines were found to be quite broad and therefore not too useful for a study of glass properties; the transition rates and lifetimes were also not much different from those found in crystalline hosts.

From a simplified point of view, a glass is a supercooled liquid. It can therefore be assumed that the spectra of RE ions in glasses will be similar to those of RE ions in liquids. The spectra in liquids show a "crystal field splitting," although with very wide lines. This is an indication that the RE ions in a liquid are surrounded by a near neighbor shell of ligands—similar to the configuration found in a solid and the same for every dissolved RE ion—and that the uncorrelated structure is only beyond the near neighbor shell. In Section 3.5 we saw that the main part of the crystal field interaction in crystals is produced by the interaction with the ligands. (This statement is not correct for the $k = 2$ term of the development of the crystal potential.) If the near neighbor coordination in a liquid is the same as in a solid, we can understand the similarity in the magnitude of the crystal field splitting of the crystal and the solution. In glasses the RE ions are incorporated as oxides. From the reasoning just cited we can expect RE spectra in glasses to be similar to those of the stable oxide modification of the particular RE ion; this expectation is verified by the experimental findings.

To exemplify the spectral properties of REs in glasses we use Eu^{3+} [RiDeSh69, We71, WeMa72], which has a well-characterized visible spectrum with very strong fluorescent transitions. It also has the $^5D_0 \rightarrow {}^7F_0$ transition, which in first order is not influenced by the crystal field, and thus permits us to estimate the minimal linewidth obtainable for spectra of REs in glasses. As mentioned, Eu ions are incorporated into the glass as Eu_2O_3, and we expect the glass to produce a spectrum similar to that of Eu_2O_3. Eu_2O_3 (like Sm_2O_3 and Gd_2O_3) is monoclinic, with space group C_{2h} and six molecules per unit cell; for the RE ion there are three different sites with symmetry C_s, each occupied by four ions.

Eu in glasses was investigated several times; here we discuss the data of Rice and De Shazer [RiDeSh69]. They compare the spectra of Eu in a borosilicate glass (40.2% SiO, 30.2% B_2O_3, 13.4% ZnO, 7.9% Na_2O, 5.4% Li_2O, 2.9% Eu_2O_3) with that of a crystal of Gd_2O_3 doped with 4% Eu, in order to get a feeling for the changes that take place in going from the crystalline to the amorphous environment.

The crystal of 4% Eu in Gd_2O_3 is preferable to pure Eu_2O_3 for this comparison because it shows no broadening of the lines due to overabsorption, or fluorescence quenching; moreover, it has roughly the same concentration of Eu as used in the glass. Absorption spectra of the glass and the Eu^{3+}:Gd_2O_3 crystal for the $^7F_0 \rightarrow {}^5D_0$ transition and an emission spectrum of the glass for the $^5D_0 \rightarrow {}^7F_0$ transition, are shown in Fig. 47, The absorption spectrum of Eu^{3+}:Gd_2O_3 shows three very narrow lines, two of them quite near to each other. Whether these three lines really correspond to ions on the three different C_s sites is an open question at this point. (The weak line at 17,175 cm^{-1} may also be a satellite (see Sec-

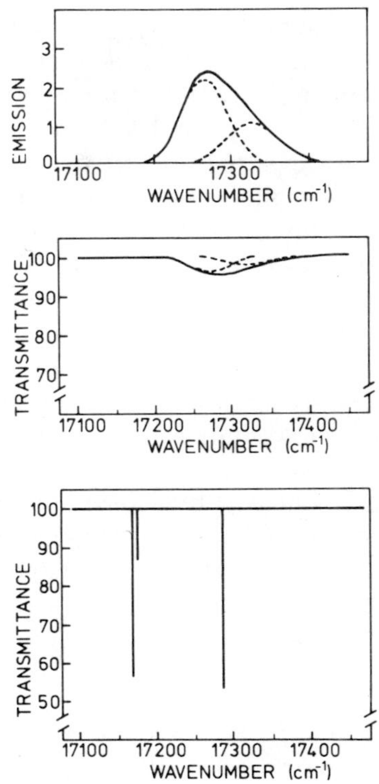

Fig. 47. Absorption and emission spectra of the $^7F_0 \rightarrow {}^5D_0$ transition in a borosilicate (BS) glass of $Eu^{3+}:Gd_2O_3$ [RiDeSh69].

Top: emission of $^5D_0 \rightarrow {}^7F_0$ transition in 3% Eu : BS glass at 4.2 K. Middle: absorption of the $^7F_0 \rightarrow {}^5D_0$ transition in 3% Eu^{3+} : BS glass at 4.2 K. Bottom: absorption of the $^7F_0 \rightarrow {}^5D_0$ transition in 4% $Eu^{3+}:Gd_2O_3$ at 4.2 K. The broken lines in the glass spectra give the decomposition into two compounds.

tion 3.17).) The two strong lines at 17,165 and 17,290 cm^{-1} are certainly due to two different sites. The emission and absorption spectra of the same transition in the glass matrix are markedly different, consisting of one broad asymmetric line more than 100 cm^{-1} wide; this asymmetric line can be decomposed into two lines, with different intensities, of approximately Gaussian shape, a width of the order of 70 cm^{-1}, and a separation of about 80 cm^{-1}, smaller than that found for the two main lines in $Eu^{3+}:Gd_2O_3$. The spectra in Fig. 47 show that though the environment near the RE ions seems to be roughly presented in going from the crystal to the glass, there is a strong inhomogeneous broadening, of the order of typical crystal field splittings, which is of course a natural consequence of the structure of a glass (random order beyond the coordinated ligands), but which also makes the spectra of RE ions in glasses less useful for structural studies.

The fluorescence spectra of the $^5D_0 \rightarrow {}^7F_1, {}^7F_2$ transitions in the same systems consist of three and five broadened lines, respectively, with average distances of the order of 100 cm^{-1}. We could naively interpret these spectra as transitions from the unsplit $^5D_0$ term to the $2J + 1$ split terms $^7F_1$ and $^7F_2$ (C_s symmetry removes the J degeneracy completely). Since the $^5D_0 \rightarrow {}^7F_0$ transition already shows a large "splitting" due to differences in the sites, this conclusion may be premature and more detailed investigations are necessary to clear up these questions.

Many measurements of the radiative and nonradiative transition probabilities of RE ions in glasses have been made. They generally produce results very much in line with the findings in crystals. In Table 42 we list the intensity parameters for Nd ions in various glasses, and compare them with the corresponding number for a typical crystal where the RE ions also see an oxygen environment, namely, Y_2O_3. Although the values for Ω_2 show some spread (as expected, because they also depend strongly on the electrostatic contribution, which is dominated by the long range environment) the Ω_4 and Ω_6 terms, which are produced predominantly by the ligands, are very similar for different glasses and for the crystal.

Since the most important application of glasses is as a host material for Nd lasers, some results for a typical material ([MaDeSh70]; another good reference in this field is [LiBu75]), a barium crown glass of composition

TABLE 42

INTENSITY PARAMETERS Ω_λ IN ALKALINE GLASSES AND Y_2O_3 FOR Nd^{3+a}

Glass composition (mole %)								
SiO_2	65	65	65	65	65	65	65	
K_2O	15	15	15	15				
MgO	20							
CaO		20						
SrO			20					
BaO				20	20	20	20	
Li_2O					15			
Na_2O						15		
K_2O							15	
Y_2O_3								100
Nd^{3+} intensity parameters ($\times 10^{+20}$ cm^2)								
Ω_2	5.1	4.6	3.8	3.7	3.8	3.8	3.7	8.6
Ω_4	4.1	3.6	3.3	2.8	3.4	3.4	2.8	5.3
Ω_6	2.8	2.9	2.7	2.3	4.1	3.5	2.3	2.9

[a] From [RiWe75, Re75].

66% SiO_2, 16% K_2O, 5% Na_2O, 5% BaO, 2% Al_2O_3, 1% Sb_2O_3, and 5% Nd_2O_3, are summarized next.

The spectra (emission and absorption) were investigated between 4.2 and 300 K. In line with our discussion of the Eu-containing glass, we expect the spectra of Nd to be similar to those of Nd_2O_3; Nd_2O_3 is trigonal with one molecule per unit cell and the two Nd ions occupy sites of C_3 point symmetry. Table 43 compares the positions of the crystal field levels of a few typical terms for the glass and Nd_2O_3 [HeMu67];

TABLE 43

ENERGY LEVELS OF TRIVALENT NEODYMIUM IONS IN GLASS AND Nd_2O_3[a]

Classification		Energy (cm^{-1})	
		Glass[b]	Nd_2O_3[c]
$^4I_{9/2}$	I	0	0
	II	55	23
	III	155	84
	IV	255	253
	V	495	496
$^4I_{11/2}$	*a*	1954	
	b	2035	
	c	2145	
	d	2220	
	e	2250	
	f	2301	
$^4F_{3/2}$	*a*	11,388	11,165
	b	11,530	11,318
$^2P_{1/2}$	*a*	23,210	22,862
$^2D_{5/2}$	*a*	23,617	23,225
	b	23,814	23,256
	c	24,241	23,558
$^2P_{3/2}$	*a*	25,967	25,687
	b	26,198	25,846

[a] The crystal field levels of the excited terms are characterized by lowercase letters, those of the ground term by Roman numerals.

[b] From [MaDeSh70].

[c] From [HeMu67].

apart from a shift in the center of gravity of the terms, there is good agreement of the size of the crystal field splitting in the two hosts. This result supports the contention that the Nd_2O_3 environment is essentially preserved in the glass. The lines themselves are much wider in the glass (up to 200 cm^{-1}) than in the crystal, which can again be attibuted to inhomogeneous broadening due to different microenvironments. The linewidth stays constant between 4.2 and 300 K, substantiating the foregoing statement. Table 44 contains a few useful parameters for the Nd-doped glass; f and A were defined earlier.

Loosely related to the glass host is ice [CoPa75]; it was used as a host for Nd ions, which were dissolved as the chloride in water. The absorption lines in this host, however, were quite narrow, indicating that the Nd ions were experiencing a long range crystalline environment. The sharpness

TABLE 44

WAVE NUMBER ($\tilde{\nu}$), HALF-WIDTH ($\Delta\tilde{\nu}$), f OSCILLATOR STRENGTH, AND A COEFFICIENT FOR SOME TRANSITIONS OF Nd IN CROWN GLASS[a]

Transition	$\tilde{\nu}$ (cm^{-1})	$\Delta\tilde{\nu}$ (cm^{-1})	f (10^{-8})	A (sec^{-1})
$^4I_{9/2} \rightarrow {}^2P_{1/2}$				
I–*a*	23,210	88	1.7	14
II–*a*	23,155	103	1.1	9
III–*a*	23,055	112	1.5	12
IV–*a*	22,955	136	0.9	7
V–*a*	22,705	166	2.6	20
$^4I_{9/2} \rightarrow {}^4F_{3/2}$				
I–*a*	11,390	84	7.7	15
II–*a*	11,335	95	20.9	41
III–*a*	11,235	120	19.7	38
IV–*a*	11,135	135	8.5	16
V–*a*	10,895	170	20.8	37
I–*b*	11,533	112	5.5	11
II–*b*	11,478	110	13.6	27
III–*b*	11,378	120	13.9	23
IV–*b*	11,278	140	4.3	8
V–*b*	11,038	170	22.4	41

[a] From [MaDeSh70]. Crystal field levels of the ground term are labeled by Roman numerals, those of the excited terms by lowercase letters.

of the lines holds out the possibility of our using them to study the structure of ice. So far, however, only preliminary results have appeared, and not even the exact site symmetry (probably a threefold symmetry) of the RE ions is known.

Another possible amorphous host is produced by means of the matrix isolation technique. In this technique a rare gas element and, e.g., an RE element in vapor form are frozen simultaneously, so that the free RE ions are imbedded in a solid environment. Such an experiment was performed on Eu in an argon matrix by Jacob *et al.* [JaMi76], and the $4f^7 6s^2 \rightarrow 4f^6 5d6s^2$ transitions were investigated. (The Eu ions are imbedded as neutral atoms.) The spectrum consists of relatively broad structures—with one zero-phonon line with phonon sidebands possibly present. So far, however, only preliminary results have been published.

We also mention an experiment that might provide interesting applications of RE spectroscopy: The incorporation of REs into biological molecules and their spectroscopy [BiGo70]. Calcium interacts with a number of proteins, but because it has closed shells it cannot be studied spectroscopically, Birnbaum *et al.* [BiGo70] showed that it can be replaced by an RE ion that has a similar ionic size (e.g., Nd^{3+}). The binding of Nd^{3+} can then be studied by the optical spectra. So far these studies have been restricted to solutions, which due to the wide lines give only limited information. But it is reasonable to expect that the RE ions will eventually be incorporated into crystals of biological molecules which will allow the spectroscopy with the sharp lines.

CHAPTER 11

Rare Earth Lasers

11.1 Introduction

A laser is a device that amplifies optical radiation by stimulated emission. The principle was first suggested by Townes [To53] for microwave radiation (maser [GoZe54]). About a decade later, Maiman [Ma60] demonstrated that it could also be made to work with optical radiation (ruby), and ever since the research in this field has been very active. Rare earth ions have been used in a number of solid state and liquid lasers; we shall describe the principle of laser action and then present the most common RE laser materials.†

One transition, the $^4F_{3/2} \rightarrow {}^4I_{11/2}$ transition (1.06 μ) for Nd^{3+} in various hosts, has received considerable attention in relation to the development of lasers. This interest was recently increased because of the potential use of Nd lasers for achieving controlled thermonuclear fusion.‡ In this book we shall not deal with applications of lasers nor with the technology of their production, since this can be found elsewhere (see, e.g., [Le66b, Ma68, ArSchu72]).

† A good review of RE lasers can be found in the volume by Reisfeld and Jørgensen [ReJo78].

‡ The state of the art of achieving controlled thermonuclear reactions by laser heating has recently been reviewed by Stickly [Sti78].

A laser consists of an optical material containing laser active ions inside a resonator, which for all practical purposes consists of two mirrors. The medium is pumped (in the cases under consideration here, mostly by light) and spontaneous radiation is emitted, which is amplified by stimulated radiation. In the resonator the radiation forms standing waves, and the different forms in which these standing waves oscillate in a particular resonator are called modes. The spontaneous emission populates a very large number of modes. The optical resonator provides feedback for only a very few modes, so that a considerable amount of energy is put into these selected modes. This concentration of energy into a few modes results in the characteristically great coherence of laser radiation.

11.2 Principles of Laser Action

We shall here outline the principle of laser operation [ScTo58, YaGo63, Go68] in order to make this book reasonably self-contained [Ma68]. A few equations from radiation theory are needed.

In a black box with walls of temperature T the radiation distribution is given by Planck's law:

$$D(\nu) = (8\pi\nu^2/c^3)[h\nu/(\exp(h\nu/k_B T) - 1)] \tag{11.1}$$

Let us assume that in this radiation field we have atoms with two non-degenerate energy levels, the ground state W_i and an excited state $W_j(W_j > W_i)$, where $W_j - W_i = h\nu$. The radiation field can cause transitions between these two energy levels according to the formula

$$D(\nu)N_i B_{ij} = D(\nu)N_j B_{ji} + N_j A_{ji} \tag{11.2}$$

where N_i and N_j are the number of atoms in states W_i and W_j and where B and A are the Einstein coefficients. The left-hand side gives the absorption from W_i to W_j, and the right-hand side gives the induced emission and the spontaneous emission. With $N_i/N_j = \exp((W_j - W_i)/k_B T)$ we find

$$D(\nu) = A_{ji}/[B_{ij}\exp((W_j - W_i)/k_B T) - B_{ji}] \tag{11.3}$$

which yields

$$B_{ji} = B_{ij} \tag{11.4a}$$

$$A_{ji} = (8\pi h\nu^3 B_{ji})/c^3 \tag{11.4b}$$

and the oscillator strength

$$f_{ji} = (mh\nu/\pi e^2)B_{ji} \tag{11.4c}$$

In a laser, atoms are pumped (ignore for the moment by what means) from state W_i to state W_j. The spontaneous emission creates a radiation field and produces induced emission. For this emission to be amplified, the gain that can be achieved in the pumped medium must be at least as large as the losses in the resonator. Thus we want to calculate this gain. The radiation field in the resonator may have frequency ν_0, where there is not only a fixed frequency ν_0, but also a frequency distribution, e.g., a Lorentzian distribution

$$g(\nu) = \frac{\Delta\nu/2\pi}{(\nu - \nu_0)^2 + (\Delta\nu/2)^2} \tag{11.5}$$

where $\Delta\nu$ is the half width of the frequency distribution. The radiation field has electrical field strength E. The change of occupation number of W_j is given by

$$dN_j/dt = N_j(E^2/8\pi)B_{ji}g(\nu) \tag{11.6}$$

where $E^2/8\pi$ is the energy density of the radiation field. Each induced transition adds $h\nu$ of energy to the radiation field; the change in energy density is given by

$$\frac{d}{dt}\left(\frac{E^2}{8\pi}\right) = h\nu\frac{dN_j}{dt} = h\nu N_j \frac{E^2}{8\pi} B_{ji}g(\nu) = \gamma(\nu)\frac{E^2}{8\pi} \tag{11.7}$$

with

$$\gamma(\nu) = h\nu N_j B_{ji} g(\nu) \tag{11.7a}$$

where $\gamma(\nu)$ is called the gain coefficient because it describes the change of energy density with respect to time. The radiation field also produces transitions from i to j, lowering the gain coefficient and yielding

$$\gamma(\nu) = h\nu(N_j B_{ji} - N_i B_{ij})g(\nu) = (\pi e^2/m)f_{ij}(N_j - N_i)g(\nu) \tag{11.8a}$$

In order to obtain the gain coefficient per unit length we divide (11.8a) by c

$$\kappa(\nu) = (\pi e^2/mc)f_{ij}(N_j - N_i)g(\nu) \tag{11.8b}$$

For laser operation to occur, the gain thus achieved must be larger than or at least equal to the loss in the configuration. The losses are produced by scattering and absorption in the laser medium, absorption in the cavity mirrors, and the extracted light (in which we are interested, in order to work with it). The loss is described by a single parameter: either $\exp(-l)$, where l is the fractional loss for one complete round trip optical pass L in the resonator (L equals twice the length of the resonator), or a time constant, which gives the field strength decay in the radiation field with respect to time, $E = E_0 \exp(-t/\tau_c)$, where [ChGe72]

$$\tau_c = -L/c \cdot \ln(1 - l) \approx L/cl \tag{11.9}$$

In order to maintain laser action, the gain must at least offset the loss; this gives the condition for laser action

$$\kappa L \geq 1$$

$$\Delta N = (N_j - N_i) \geq \frac{l}{L} \frac{mc}{\pi e^2 f_{ji} g(\nu_0)}$$

$$\Delta N \geq \frac{l}{L} \frac{4\pi^2 \nu^2 \Delta\nu \tau_R}{c^2} = \frac{4\pi^2 \nu^2 \Delta\nu}{c^3} \frac{\tau_R}{\tau_c} \tag{11.10†}$$

where τ_R is the radiative lifetime of the excited state. If the laser medium has refractive index n, one has to replace c by c/n in eq. (11.10).

This population difference has to be maintained in order to have laser operation (if states i and j have degeneracies g_i and g_j, the population difference is written $N_j - (g_j/g_i)N_i$).

In actual operation, we distinguish between three and four level lasers. In a three level laser the system is pumped by absorption into a level W_k, where $W_k > W_j$. The energy migrates from W_k to W_j and $W_j \to W_i$ is the laser transition. In a four level laser there is an additional level W_l with $W_l < W_i$; the laser transition is $W_j \to W_i$. A four level system has a low threshold because the population W_i can be very small if $W_i - W_l$ greatly exceeds kT_r (where T_r is room temperature). Thus the widely used Nd^{3+} laser is a four level laser in which the states (W_k) above ${}^4F_{3/2}$ (W_j) are pumped, the energy is transferred into this state, and ${}^4F_{3/2} \to {}^4I_{11/2}$ ($W_j \to W_i$) is the actual laser transition (${}^4I_{11/2}$ (W_i) is 2000 cm^{-1} above the ground state ${}^4I_{9/2}$ (W_l).)

A low threshold is important for laser operation. In inspecting eq. (11.10) we see that the RE ions are useful for obtaining low thresholds, because they have levels with narrow linewidth $\Delta\nu$, which makes the right-hand side of this equation small. The loss time τ_c is a function of the host material. We also see that population inversion is more easily obtained in a four level than in a three level laser. Here again, REs are valuable because of their many energy levels, where fluorescent transitions that do not terminate in the ground term often occur.

† Here for $g(\nu_0)$ the Lorentzian distribution function (11.5) was used. Often a Gaussian may be more appropriate (e.g., in a glass laser); this introduces only slight modifications in the formulas, see [ScTo58].

11.3 Typical Rare Earth Lasers: Nd:YAG and Nd:Glasses

A summary of some RE laser material properties is given in Tables 45–48 (see [Pr71]). Laser action has been achieved in Pr^{3+}, Nd^{3+}, Eu^{3+}, Ho^{3+}, Er^{3+}, Tm^{3+}, and Yb^{3+}, and in Sm^{2+}, Dy^{2+}, and Tm^{2+}. By far the most important ion is Nd^{3+} in a glass host or in YAG (yttrium aluminium garnet). Trivalent RE ions in hosts have a number of features that make them attractive for use in lasers; they have very narrow absorption (and emission) lines; the emission quantum efficiency is near unity for states that are a couple of thousand wavenumbers above the next state; and examples can be found where the final state is at an energy considerably above the ground state. All these features make the required inversion population ΔN (eq. (11.10)) low. The main drawback of trivalent RE ions is that they cannot be pumped very effectively, because they absorb only in the very narrow RE lines with their low oscillator strength. This disadvantage can sometimes be overcome if the system is doped with another RE ion that transfers its energy to the lasing ion. However, the doped RE ion also has only a small absorption cross section. In principle, divalent RE ions (for which the very strong $4f^N \rightarrow 4f^{N-1}5d$ transition extends well into the optical energy region and can be used for effective pumping) do not entail the main disadvantage of the trivalent REs. The energy in these divalent REs is transferred to the 4f states, which in turn provide the laser action. So far, however, divalent ions have not been widely used because of other drawbacks. If incorporated into a suitable host (such as CaF_2), the divalent oxidation state is not the most stable one and a divalent to trivalent conversion may occur during laser operation. The divalent ions show a stronger electron–phonon interaction than their trivalent counterparts, which results in a smaller fluorescent quantum efficiency and in strong vibronic sidebands. Finally, because of the large width of the 5d states, absorption from the lasing level into the 5d states may occur, thereby lowering the population inversion.

There is an extensive literature on lasers and RE lasers [Le66b, Ma68, ArSchu72, Pr71].

From Tables 45–47 we see that in all the cases mentioned the laser-active ion is present only as an impurity in a host crystal. Some RE compounds, however, have the appealing property of showing laser action in the concentrated form of the crystal (see Table 48 for some examples). This property brings about much higher concentrations of laser active ions, which in principle makes possible the construction of

TABLE 45

LASER TRANSITIONS FOR RE IONS[a,b]

Ion	λ (μm)	Transition	Terminal level (cm^{-1})	Host	Reference
Pr^{3+}	1.05	$^1G_4 \rightarrow {}^3H_4$		$CaWO_4$	[YaPo62]
				$CaNb_2O_6$	[BaPo63]
	0.60	$^3P_0 \rightarrow {}^3H_6$	4200	LaF_3	[SoMu63]
Nd^{3+}	1.06	$^4F_{3/2} \rightarrow {}^4I_{11/2}$	2000	CaF_2	[Jo63]
				BaF_2	[Jo63]
				Y_2O_3	[HoSo64]
				$Y_3Ga_5O_{12}$	[GeMa64]
				Glass	[Sn61]
				$CaWO_4$	[JoBo62]
				$Y_3Al_5O_{12}$	[GeMa64]
				LaF_3	[Jo63]
				SrF_2	[Jo63]
				$LiNbO_3$	[EvKo67]
Nd^{3+}	1.34	$^4F_{3/2} \rightarrow {}^4I_{13/2}$	4000	$CaWO_4$	[JoTh63]
				Glass	[Ma63a]
				$Y_3Al_5O_{12}$	[DeRö68]
Nd^{3+}	0.91	$^4F_{3/2} \rightarrow {}^4I_{9/2}$		$CaWO_4$	[Jo63]
				Glass	[Ma63b]
Eu^{3+}	0.61	$^5D_0 \rightarrow {}^7F_2$	860	Y_2O_3	[Ch63]
Ho^{3+}	2.05	$^5I_7 \rightarrow {}^5I_8$		$CaWO_4$	[Jo63]
				CaF_2	[Jo63]
				$Y_3Al_5O_{12}$	[JoGe65]
				Glass	[GaGi62]
				$Y_3Fe_3O_{12}$	[StaMa76]
Er^{3+}	1.61	$^4I_{13/2} \rightarrow {}^4I_{15/2}$		$CaWO_4$	[KiDu62a]
				Glass	[SnWo65]
				$Y_3Al_5O_{12}$	[JoGe65]
Tm^{3+}	1.91	$^3H_4 \rightarrow {}^3H_6$		$CaWO_4$	[Jo63]
				CaF_2	[Jo63]
				Glass	[GaGi67]
				$Y_3Al_5O_{12}$	[JoGe65]
Yb^{3+}	1.03	$^2F_{5/2} \rightarrow {}^2F_{7/2}$		$Y_3Al_5O_{12}$	[JoGe65]
				Glass	[EtGa62]
Sm^{2+}	0.71	$5d \rightarrow {}^7F_1$	260	CaF_2	[KaGa61a]
		$^5D_0 \rightarrow {}^7F_1$		SrF_2	[WoKa62]
Dy^{2+}	2.36	$^5I_7 \rightarrow {}^5I_8$		CaF_2	[KiDu62b]
Tm^{2+}	1.12	$^2F_{5/2} \rightarrow {}^2F_{7/2}$		CaF_2	[KiDu62a]

[a] For additional data see [ChGe72, Jo66].
[b] From [ChGe72].

TABLE 46

CONTINUOUS WAVE LASER TRANSITION[a] OBSERVED AT ROOM TEMPERATURE[b]

Ion	λ (μm)	Transition	Terminal level energy (cm^{-1})	Host	Reference
Nd^{3+}	1.0582	$^4F_{3/2} \rightarrow {}^4I_{11/2}$	~2000	$CaWO_4$	[JoBo62]
Nd^{3+}	1.06			Glass	[Yo63]
Nd^{3+}	1.0795			$YAlO_3$	[BaKa69]
Nd^{3+}	1.319	$^4F_{3/2} \rightarrow {}^4I_{13/2}$		$Y_3Al_5O_{12}$	[Sm68]

[a] Many solid state lasers incorporate a flash lamp that pumps the laser every few seconds. In a few cases operation under continuous illumination (continuous wave operation) is achieved, and some transitions for which this is the case are summarized in this table.

[b] From [ChGe72].

very small lasers. So far, however, large-scale application of these materials has been hampered by other drawbacks.

We conclude with a few remarks on two Nd lasers [ArSchu72]. The Nd glass laser [SnYo68, ChGe72] has special importance because it provides the highest power and shortest pulses of any laser known. The

TABLE 47

CONTINUOUS WAVE LASER TRANSITIONS[a] OBSERVED ONLY BELOW ROOM TEMPERATURE[b]

Ion	λ (μm)	Transition	Terminal level (cm^{-1})	Host	Reference
Dy^{2+}	2.36	$^5I_7 \rightarrow {}^5I_8$	30	CaF_2	[Ya62]
Ho^{3+}	2.12		518	$Y_3Al_5O_{12}$	[JoGe65]
Ho^{3+}	2.10		462	$Y_3Al_5O_{12} + Cr^{3+}$	[JoGe65]
Ho^{3+}	2.10		462	$Er_{1.48}Y_{1.5}Al_5O_{12}$	[JoGe65]
Tm^{3+}	2.01	$^3H_4 \rightarrow {}^3H_6$	582	$Y_3Al_5O_{12}$	[JoGe65]
Tm^{3+}	2.01		582	$Y_3Al_5O_{12} + Cr^{3+}$	[JoGe65]
Tm^{3+}	2.01		582	$Er_{1.48}Y_{1.5}Al_5O_{12}$	[JoGe65]
Tm^{3+}	1.93		0	Er_2O_3	[SoHo65]
Tm^{2+}	1.12	$^2F_{5/2} \rightarrow {}^2F_{7/2}$	0	CaF_2	[DuKi63]

[a] Many solid state lasers incorporate a flash lamp that pumps the laser every few seconds. In a few cases operation under continuous illumination (continuous wave operation) is achieved, and some transitions for which this is the case are summarized in this table.

[b] From [ChGe72].

TABLE 48

LASER TRANSITIONS IN UNDILUTED RE SALTS

Ion	λ (μm)	Transition	Terminal level (cm^{-1})	Crystal	Temperature	Reference
Pr^{3+}	0.645	$^3P_0 \rightarrow {}^3F_2$	4850	$PrCl_3$	300 K	[Va71]
Pr^{3+}	0.645	$^3P_0 \rightarrow {}^3F_2$	4850	$PrBr_3$	300 K	[Va71]
Pr^{3+}	0.645	$^3P_0 \rightarrow {}^3F_2$	4850	$PrCl_3$	300 K	[GeKi73]
Pr^{3+}	0.616	$^3P_0 \rightarrow {}^3H_6$	4230	$PrCl_3$	65 K	[GeKi73]
Pr^{3+}	0.530	$^3P_0 \rightarrow {}^3H_4$	2137	$PrCl_3$	35 K	[GeKi73]
Nd^{3+}	1.05	$^4F_{3/2} \rightarrow {}^4I_{11/2}$	2000	NdP_5O_{14}	300 K	[WeDa73]
Ho^{3+}	2	$^5I_7 \rightarrow {}^5I_8$	0	HoF_3	77 K	[DeSo71]

advantages of glass are its optical isotropy and its durability; moreover, it can be doped both heavily and uniformly, and can easily be fabricated in large sizes (up to 2 m in length with a diameter of close to 0.1 m). The optical properties can be varied over quite a wide range. The linewidth in this system is large as a consequence of inhomogeneous broadening. This makes possible the storage of large amounts of energy in a system that is operated in a Q-switched mode. This laser operation technique produces very intense bursts of radiation. During the pumping process, the quality (Q) of the resonator is kept low (small τ_c in eq. (11.10)), which makes it possible to build up a large population difference. The quality of the resonator is then increased, which results in a very intense laser operation. The main disadvantage of a glass laser is its low thermal conductivity, which limits the repetition rate.

The Nd^{3+}:YAG laser is of special importance because it shows continuous wave (CW) operation at room temperature. (Many solid state lasers can be operated only in the pulsed mode: A flash lamp produces the inversion population necessary for laser operation whereupon laser activity is observed. Laser operation achieved by continuous illumination of the active material, is called CW operation; the power output in large CW systems approaches the kilowatt region.) YAG has very attractive mechanical properties and high thermal conductivity; it is available in very good optical quality, so that rods 1 cm in diameter and up to 15 cm long are possible. The lifetime of the $^4F_{3/2}$ state in this host is 0.23 msec, the quantum efficiency for fluorescence is unity, and the linewidth of the fluorescing state is 180 GHz.

References

[AbBl70] A. Abragam and B. Bleaney, "Electron Paramagnetic Resonance of Transition Ions." Oxford (Clarendon) Univ. Press, London and New York, 1970.

[AbKu66] I. D. Abella, N. A. Kurnit, and S. R. Hartmann, *Phys. Rev.* **141**, 391 (1966).

[AlDe73] M. Altarelli and D. L. Dexter, *Phys. Rev. B* **7**, 5335 (1973).

[AlFi75] M. Alonso and E. J. Finn, "University Physics." Addison-Wesley, Reading, Massachusetts, 1975.

[An58] P. W. Anderson, *Phys. Rev.* **109**, 1492 (1958).

[ArSchu72] F. T. Arecchi and E. O. Schulz-Dubois, eds. "Laser Handbook." North-Holland Publ., Amsterdam, 1972.

[Au76] F. Auzel, *in Proc. 1975 Int. Conf. Luminesc.* p. 715. North-Holland Publ., Amsterdam, 1976.

[AxBu66] J. D. Axe and G. Burns, *Phys. Rev.* **152**, 331 (1966).

[AxDi62] J. D. Axe and G. H. Dieke, *J. Chem. Phys.* **37**, 2364 (1962).

[AxSo63] J. D. Axe and P. P. Sorokin, *Phys. Rev.* **130**, 945 (1963).

[Ba62] C. J. Ballhausen, "Introduction to Ligand Field Theory." McGraw-Hill, New York, 1962.

[Ba71] J. M. Baker, *Rep. Progr. Phys.* **34**, 109 (1971).

[Ba75] R. Baumert, Thesis, Berlin (1975).

[BaBl58] J. M. Baker and B. Bleaney, *Proc. Roy. Soc.* **A245**, 156 (1958).

[BaHu63] M. Ball, M. T. Hutchings, M. J. M. Leask, and W. P. Wolf, *Proc. Int. Conf. Low Temp. Phys., 8th, London, 1962* (R. O. Davies, ed.). Butterworths, London, 1963.

[BaKa69] Kh. S. Bagdasarov, A. A. Kaminiskii, and B. Sobolev, *Sov. Phys. Crystallogr.* **13**, 779 (1969).

[BaKa75a] J. E. Battison, A. Kasten, M. J. M. Leask, J. B. Lowry, and K. J. Maxwell, *J. Phys. C Solid State Phys.* **9**, 1345 (1975).

[BaKa75b] B. Batlogg, E. Kaldis, A. Schlegel, and P. Wachter, *Phys. Rev. B* **12**, 3940 (1975).

[BaKa76a] B. Batlogg, E. Kaldis, A. Schlegel, and P. Wachter, *Phys. Rev. B* **14**, 5503 (1976).

[BaKa76b] B. Batlogg, E. Kaldis, A. Schlegel, G. von Schulthess, and P. Wachter, *Solid State Commun.* **19**, 673 (1976).

[BaPe75] R. Baumert, J. Pelzl, and S. Hüfner, *Solid State Commun.* **16**, 345 (1975).

[BaPo63] A. A. Ballmann, S. P. S. Porto, and A. Yariv, *J. Appl. Phys.* **34**, 3155 (1963).

[BaScha70] E. Bayer and G. Schaack, *Phys. Status Solidi* **41**, 827 (1970).

[BaVe69] D. M. S. Bagguley and G. Vella-Coleiro, *J. Phys. C* **2**, 2310 (1969).

[Be08] J. Becquerel, *Phys. Z.* **8**, 632 (1908).

[Be29] H. A. Bethe, *Ann. Phys.* **3**, 133 (1929).

[Be63] E. F. Bertaut, "Magnetism" (G. T. Rado and H. Suhl, eds.), Vol. III, p. 149. Academic Press, New York, 1963.

[Be71] P. J. Becker, *Phys. Status Solidi* (*b*) **43**, 583 (1971).

[Be76] P. J. Becker, *Phys. Status Solidi* (*b*) **74**, 285 (1976).

[BeGe75] P. J. Becker and G. A. Gehring, *Solid State Commun.* **16**, 795 (1975).

[BeKa73a] W. Berkhahn, H. G. Kahle, L. Klein, and H. C. Schopper, *Phys. Status solidi* (*b*) **55**, 265 (1973).

[BeKa73b] W. Becher and H. Kalbfleisch, *Phys. Status Solidi* (*b*) **60**, 557 (1973).

[BeLa71] P. J. Becker and J. Laugsch, *Phys. Status Solidi* (*b*) **44**, K109 (1971).

[BeLa73] P. J. Becker and J. Laugsch, *Proc. Int. Conf. Magn., Moscow* (1973).

[BeLe70] E. Belorizky, M. J. M. Leask, and K. J. Maxwell, *Colloq. Eléments Terres Rares* p. 453 (1970).

[BeLe72a] P. J. Becker, M. J. M. Leask, K. J. Maxwell, and R. N. Tyte, *Phys. Lett.* **41A**, 205 (1972).

[BeLe72b] P. J. Becker, M. J. M. Leask, and R. N. Tyte, *J. Phys. C: Solid State Phys.* **5**, 2027 (1972).

[BeMa67] E. F. Bertaut and J. Mareschal, *Solid State Commun.* **5**, 93 (1967).

[Bi67] R. J. Birgeneau, *Phys. Rev. Lett.* **19**, 160 (1967).

[BiGo70] E. R. Birnbaum, J. E. Gomez, and D. W. Darnall, *J. Am. Chem. Soc.* **92**, 5287 (1970).

[BiMa67] H. J. Bielen, J. Mareschal, and J. Sivardière, *Z. Angew. Phys.* **23**, 243 (1967).

[BiNe68] S. S. Bishton and D. J. Newman, *J. Phys. Chem. Solids* **29**, 1245 (1968).

[BiNe70] S. S. Bishton and D. J. Newman, *J. Phys. C, Solid State Phys.* **3**, 1753 (1970).

[BiZv71] V. I. Bilak, G. M. Zverev, G. O. Karapetyan, and A. M. Onishchenko, *JETP Lett.* **14**, 199 (1971).

[Bl59] N. Bloemberger, *Phys. Rev. Lett.* **2**, 84 (1959).

[Bl65] M. Blume, *Phys. Rev.* **174**, 35 (1965).

[Bl73] B. Bleaney, *Physica* **69**, 317 (1973).

[Bl76] G. Blasse, *Structure and Bonding* **26**, 43 (1976).

[BlEl51] B. Bleaney, R. J. Elliott, and H. E. D. Scovil, *Proc. Phys. Soc.* (*London*) **A64**, 933 (1951).

[BlSh66] J. Blok and D. A. Shirley, *Phys. Rev.* **143**, 278 (1966).

[BoDu52] G. S. Bogle, H. J. Duffus, and H. E. D. Scovil, *Proc. Phys. Soc.* (*London*) **A65**, 760 (1952).

[BoGr73] D. Boal, P. Grünberg, and J. A. Koningstein, *Phys. Rev. B* **7**, 4757 (1973).

[Br65] W. E. Bron, *Phys. Rev.* **140**, A2005 (1965).

[BrGo45] J. F. Broer, C. J. Gorter, and J. Hoogschagen, *Physica* **11**, 231 (1945).

[BrNe67] M. I. Bradbury and D. J. Newman, *Chem. Phys. Lett.* **1**, 44 (1967).

[Bu62] H. A. Buckmaster, *Can. J. Phys.* **40**, 1670 (1962).

[Bu69] G. Burns, *Solid State Commun.* **7**, 1073 (1969).

[BuDo71] W. Buxmeyer, I. N. Douglas, A. Eren, H. Heischmann, K. H. Hellwege, and S. Leutloff, *Phys. Status Solidi* (*b*) **47**, K33 (1971).
[BuJu62] G. Busch, P. Junod, M. Risi, and O. Vogt, *Proc. Int. Conf. Semicond., Exeter* p. 727 (1962).
[BuJu64] G. Busch, P. Junod, and P. Wachter, *Phys. Lett.* **12**, 11 (1964).
[BuWa66] G. Busch and P. Wachter, *Phys. Kondens. Mater.* **5**, 232 (1966).
[BuWi67] R. A. Buchanan, K. A. Wickersheim, J. J. Pearson, and G. F. Herrmann, *Phys. Rev.* **159**, 245, 251 (1967).
[CaCr76] W. T. Carnall, H. Crosswhite, H. M. Crosswhite, and J. G. Conway, *J. Chem. Phys.* **64**, 3582 (1976).
[CaCr77] W. T. Carnall, H. Crosswhite, and H. M. Crosswhite, Energy Level Structure and Transition Probabilities of Trivalent Lanthanides in LaF_3. Argonne Nat. Lab. Rep. (1977).
[CaFi68a] W. T. Carnall, P. R. Fields, and K. Rajnak, *J. Chem. Phys.* **49**, 4412 (1968).
[CaFi68b] W. T. Carnall, P. R. Fields, and K. Rajnak, *J. Chem. Phys.* **49**, 4424, 4443, 4447, 4450 (1968).
[Ch63] N. C. Chang, *J. Appl. Phys.* **34**, 3500 (1963).
[ChGe72] R. B. Chesler and J. E. Geusic, *in* "Laser Handbook" (F. T. Arecchi and E. O. Schulz-Dubois, eds.), Vol. 1, p. 325. North-Holland Publ., Amsterdam, 1972.
[ChHa76] Y. C. Chen and S. R. Hartmann, *Phys. Lett.* **58A**, 201 (1976).
[ClHu71] R. H. Clarke and C. A. Hutchison, Jr., *Phys. Rev. Lett.* **27**, 638 (1971).
[ClWo68] R. B. Clover and W. P. Wolf, *Solid State Commun.* **6**, 331 (1968).
[Co64] R. L. Cohen, *Phys. Rev.* **134**, A94 (1964).
[Co75a] R. Courths, *Z. Phys.* **B22**, 253 (1975).
[Co75b] P. A. Cox, *Structure and Bonding* **24**, 59 (1975).
[CoBa78] G. M. Copland, G. Balasubramanian, and D. J. Newman, *J. Phys. C, Solid State Phys.* **11**, 2029 (1978).
[CoCo63] R. A. Cowley, W. Cochran, B. N. Brockhouse, and A. D. B. Woods, *Phys. Rev.* **131**, 1030 (1963).
[CoEl70] A. H. Cooke *et al.*, *Solid State Commun.* **8**, 689 (1970).
[CoGe65] A. H. Cooke, K. A. Gehring, M. J. M. Leask, D. Smith, and J. H. M. Thornley, *Phys. Rev. Lett.* **14**, 685 (1965).
[CoGr76] R. D. Cowan and D. C. Griffin, *J. Opt. Soc. Am.* **66**, 1010 (1976).
[CoHü75] R. Courths and S. Hüfner, *Z. Phys.* **B22**, 245 (1975).
[CoMa71] A. H. Cooke, D. M. Martin, and M. R. Wells, *Solid State Commun.* **9**, 519 (1971).
[CoMa72] J. H. Colwell and B. W. Mangum, *Solid State Commun.* **11**, 83 (1972).
[CoMe75] R. L. Cone and R. S. Meltzer, *J. Chem. Phys.* **62**, 3573 (1975).
[CoMo67a] E. Cohen and H. W. Moos, *Phys. Rev.* **161**, 258 (1967).
[CoMo67b] E. Cohen and H. W. Moos, *Phys. Rev.* **161**, 268 (1967).
[CoPa75] L. Couture and A. Le Paillier-Malécot, *J. Phys. C Solid State Phys.* **8**, L67 (1975).
[CoRi68] E. Cohen, L. A. Riseberg, and H. W. Moos, *Phys. Rev.* **175**, 521 (1968).
[CoSh35] E. U. Condon and G. H. Shortley, "The Theory of Atomic Spectra." Cambridge Univ. Press, London and New York, 1953.
[CoSw72] A. H. Cooke, S. J. Swithenby, and M. R. Wells, *Solid State Commun.* **10**, 265 (1972).
[CoSw73] A. H. Cooke, S. J. Swithenby, and M. R. Wells, *J. Phys. C, Solid State Phys.* **6**, 2209 (1973).

[Cr77] H. M. Crosswhite, in "Spectroscopie des Éléments de Transition et des Éléments Lourds dans les Solides," p. 65. Editions du CNRS, 1977.

[CrCr68] H. Crosswhite, H. M. Crosswhite, and B. R. Judd, *Phys. Rev.* **174**, 89 (1968).

[CrCr76] H. M. Crosswhite, H. Crosswhite, F. W. Kaseta, and R. Sarup, *J. Chem. Phys.* **64**, 1981 (1976).

[CrDi61] H. M. Crosswhite and G. H. Dieke, *J. Chem. Phys.* **35**, 1535 (1961).

[CrMo67] H. M. Crosswhite and H. W. Moos, *Conf. Opt. Properties Ions in Crystals*, p. 3. Wiley (Interscience), New York, 1967.

[CuNe69] M. M. Curtis, D. J. Newman, and G. E. Stedman, *J. Chem. Phys.* **50**, 1077 (1969).

[CuNe70] M. M. Curtis and D. J. Newman, *J. Chem. Phys.* **52**, 1340 (1970).

[De53] D. L. Dexter, *J. Chem. Phys.* **21**, 836 (1953).

[De62] D. L. Dexter, *Phys. Rev.* **126**, 1962 (1962).

[deGe58] P. G. de Gennes, *J. Phys. Chem. Solids* **7**, 345 (1958).

[DePe77] C. Delsart, N. Pelletier-Allard, and R. Pelletier, *Phys. Rev. B* **16**, 154 (1977).

[DeRö68] U. Deserno, D. Röss, and G. Zeidler, *Phys. Lett.* **28A**, 422 (1968).

[DeShDi63] L. G. DeShazer and G. H. Dieke, *J. Chem. Phys.* **38**, 2190 (1963).

[DeSo71] D. P. Devor, B. H. Soffer, and M. Robinson, *Appl. Phys. Lett.* **18**, 122 (1971).

[Di68] G. H. Dieke, "Spectra and Energy Levels of Rare Earth Ions in Crystals." Wiley, New York, 1968.

[DiBa68] B. Di Bartolo, "Optical Interactions in Solids." Wiley, New York, 1968.

[DiBa75] B. Di Bartolo (ed.), "Optical Properties of Ions in Solids." Plenum Press, New York, 1975.

[DiCr63] G. H. Dieke and H. M. Crosswhite, *Appl. Opt.* **2**, 675 (1963).

[DiCr67] G. H. Dieke and H. M. Crosswhite, *Physica* **33**, 212 (1967).

[DiHa70] J. O. Dimmock, J. Hanus, and J. Feinleib, *J. Appl. Phys.* **41**, 1088 (1970).

[DiOv58] B. J. Dick and A. W. Overhauser, *Phys. Rev.* **112**, 90 (1958).

[DiPa64] G. H. Dieke and B. Pandey, *J. Chem. Phys.* **41**, 1952 (1964).

[DiWa61] J. F. Dillon and L. R. Walker, *Phys. Rev.* **124**, 1401 (1961).

[Do71] I. N. Douglas, *Phys. Status Solidi (b)* **48**, 327 (1971).

[DuKi63] R. C. Duncan, Jr. and Z. J. Kiss, *Appl. Phys. Lett.* **3**, 23 (1963).

[EaHo69] D. E. Eastman, F. Holtzberg, and S. Methfessel, *Phys. Rev. Lett.* **23**, 226 (1969).

[Ed60] A. R. Edmonds, "Angular Momentum in Quantum Mechanics." Princeton Univ. Press, Princeton, New Jersey, 1960.

[Ei63a] J. C. Eisenstein, *J. Chem. Phys.* **39**, 2128 (1963).

[Ei63b] J. C. Eisenstein, *J. Chem. Phys.* **39**, 2134 (1963).

[EiCo69] M. Eibschütz, R. L. Cohen, and K. W. West, *Phys. Rev.* **178**, 572 (1969).

[EiHo70] M. Eibschütz, L. Holmes, J. P. Maita, and L. G. Van Uitert, *Solid State Commun.* **8**, 1815 (1970).

[El72] R. J. Elliott (ed.), "Magnetic Properties of Rare Earth Metals." Plenum Press, New York, 1972.

[ElGe71] R. J. Elliot, G. A. Gehring, A. P. Malozemoff, S. R. P. Smith, W. S. Staude, and R. N. Tyte, *J. Phys. C Solid State Phys.* **4**, L179 (1971).

[ElNe67] M. M. Ellis and D. J. Newman, *J. Chem. Phys.* **47**, 1986 (1967).

[ElNe69] M. M. Ellis and D. J. Newman, *J. Chem. Phys.* **49**, 4037 (1969).

[ElSt52] R. J. Elliott and K. W. H. Stevens, *Proc. Roy. Soc. London* **A215**, 437 (1952).

[ElSt53a] R. J. Elliott and K. W. H. Stevens, *Proc. Roy. Soc. London* **A218**, 553 (1953).

[ElSt53b] R. J. Elliott and K. W. H. Stevens, *Proc. Roy. Soc. London* **A219**, 387 (1953).

[Er61] E. E. Erath, *J. Chem. Phys.* **34**, 1985 (1961).

[Er75] L. E. Erickson, *Opt. Commun.* **15**, 246 (1975).

[Er77] L. E. Erickson, *Phys. Rev. B*16, 4731 (1977).

[EtGa62] H. W. Ethel, H. W. Gandy, and R. J. Ginther, *Appl. Opt.* **1**, 534 (1962); see also E. Snitzer and G. Young, *in* "Lasers" (A. K. Levine, ed.), Vol. 2. Dekker, New York, 1968.

[Ev64] R. C. Evans, "An Introduction to Crystal Chemistry." Cambridge Univ. Press, London and New York, 1964.

[EvKo67] N. Evlanova, A. Kovalev, V. Koptisik, L. Kovnienko, A. Prokhorov, and L. Rashkovich, *JETP Lett.* **5**, 291 (1967).

[Fa69] R. Faulhaber, Thesis, Darmstadt (1968); see also R. Faulhaber and S. Hüfner, *Solid State Comm.* **7**, 389 (1969); *Z. Phys.* **228**, 235 (1969).

[FaRa59] U. Fano and G. Racah, "Irreducible Tensorial Sets." Academic Press, New York, 1959.

[Fe74] J. Ferré, *J. Phys.* **35**, 781 (1974).

[FeEb64] J. Ferch, W. Ebenhöhl, and V. J. Ehlers, *Phys. Verhandl. DPG* **4**, 250 (1964).

[FeSc69] J. Feinleib, W. J. Scouler, J. O. Dimmock, J. Hanus, T. B. Reed, and C. R. Pidgeon, *Phys. Rev. Lett.* **22**, 1385 (1969).

[FeTi65] P. P. Feofilov, V. A. Timofeeva, M. H. Tolstoi, and L. M. Belyaev, *Opt. Spectrosc.* **19**, 451 (1965).

[FeTr69] P. P. Feofilov and A. K. Trofimov, *Opt. Spectrosc.* **27**, 291 (1969).

[FiJo57] E. Fick and G. Joos, Kristallspektren, *in* "Handbuch der Physik" (S. Flügge, ed.), Vol. 28, p. 205. Springer-Verlag, Berlin and New York, 1957.

[FiOr61] C. B. P. Finn, R. Orbach, and W. P. Wolf, *Proc. Phys. Soc.* **77**, 261 (1961).

[FiRu59] D. R. Fitzwater and R. E. Rundle, *Z. Kristallogr.* **112**, 362 (1959).

[FiSi63] D. B. Fitchen, R. H. Silsbee, T. A. Fulton, and E. L. Wolf, *Phys. Rev. Lett.* **11**, 275 (1963).

[FlHa75] R. Flach, D. S. Hamilton, P. M. Selzer, and W. M. Yen, *Phys. Rev. Lett.* **35**, 1034 (1975).

[Fö48] T. Förster, *Ann. Phys.* **2**, 55 (1948).

[Fö65] T. Förster, *in* "Modern Quantum Chemistry" (O. Sinanoglu, ed.), Part III, Chapter B.1. Academic Press, New York, 1965.

[FrHo68] M. J. Freiser, F. Holtzberg, S. Methfessel, G. D. Pettit, M. W. Shafer, and J. C. Suits, *Helv. Phys. Acta* **41**, 832 (1968).

[FrMe68] M. J. Freiser, S. Methfessel, and F. Holtzberg, *J. Appl. Phys.* **39**, 900 (1968).

[FrSha76] U. Fritzler and G. Schaack, *J. Phys. C Solid State Phys.* **9**, L23 (1976).

[FrWa62] A. J. Freeman and R. E. Watson, *Phys. Rev.* **127**, 2058 (1962).

[FrWa63] A. J. Freeman and R. E. Watson, *Phys. Rev.* **132**, 706 (1963).

[FrWa65] A. J. Freeman and R. E. Watson, *Phys. Rev.* **139**, A1606 (1965).

[GaGi62] H. W. Gandy and R. J. Ginther, *Proc. IRE* **50**, 2113 (1962); see also E. Snitzer and G. Young, *in* "Lasers" (A. K. Levine, ed.), Vol. 2. Dekker, New York, 1968.

[GaGi67] H. W. Gandy, R. J. Ginther, J. F. Weller, *J. Appl. Phys.* **38**, 3030 (1967).

[GeGe75] G. A. Gehring and K. A. Gehring, *Rep. Progr. Phys.* **38**, 1 (1975).

[GeGi57] S. Geller and M. A. Gilleo, *J. Phys. Chem. Solids* **3**, 30 (1957).

[GeKi73] K. German, A. Kiel, and H. J. Guggenheim, *Appl. Phys. Lett.* **22**, 87 (1973).

[GeLe69] K. A. Gehring, M. J. M. Leask, and J. H. M. Thornley, *J. Phys. C, Solid State Phys.* **2**, 484 (1969).

[GeMa64] J. E. Geusic, H. M. Marcos, and L. G. Van Uitert, *Appl. Phys. Lett.* **4**, 182 (1964).

[GeMa76] A. Z. Genack, R. M. Macfarlane, and R. G. Brewer, *Phys. Rev. Lett.* **37**, 1078 (1976).

[GeMo72] G. A. Gehring, A. P. Malozemoff, W. Staude, and R. N. Tyte, *J. Phys. Chem. Solids* **33**, 1487 (1972).

[Go68] J. P. Gordon, *in* "Laser Technology and Applications" (S. L. Marshall, ed.), p. 27. McGraw-Hill, New York, 1968.

[GoSc65] L. S. Goodman and K. Schlüpmann, *Z. Phys.* **178**, 235 (1965).

[GoZe54] J. P. Gordon, H. J. Zeiger, and C. H. Townes, *Phys. Rev.* **95**, 282 (1954).

[Gr63a] J. B. Gruber, *J. Chem. Phys.* **38**, 946 (1963).

[Gr63b] J. S. Griffith, *Phys. Rev.* **132**, 316 (1963).

[Gr64] J. S. Griffith, "The Theory of Transition Metal Ions." Cambridge Univ. Press, London and New York, 1964.

[GrHe61] I. Grohmann, K. H. Hellwege, and H. G. Kahle, *Z. Phys.* **164**, 243 (1961).

[GrHü67] J. Grabmaier, S. Hüfner, E. Orlich, and J. Pelzl, *Phys. Lett.* **24a**, 680 (1967).

[GrHü69a] P. Grünberg, S. Hüfner, E. Orlich, and J. Schmitt, *Phys. Rev.* **184**, 285 (1969).

[GrHü69b] P. Grünberg, S. Hüfner, E. Orlich, and J. Schmitt, *J. Appl. Phys.* **40**, 1501 (1969).

[GrKo67] D. M. Gruen, C. W. DeKock, R. L. McBeth, *Ad. Chem. Ser.* **71**, 102 (1967).

[GrSa63] J. B. Gruber and R. A. Satten, *J. Chem. Phys.* **39**, 1455 (1963).

[Gü74] G. Güntherodt, *Phys. Kondens. Mater.* **18**, 37 (1974).

[GüIm71] G. Güntherodt, D. M. Imboden, and P. Wachter, *Phys. Kondens. Mater.* **12**, 292 (1971).

[HaHa72] R. T. Harley, W. Hayes, and S. R. P. Smith, *J. Phys. C Solid State Phys.* **5**, 1501 (1972).

[HaHo75] A. Hasson, R. M. Hornreich, Y. Komet, B. M. Wanklyn, and I. Yaeger, *Phys. Rev. B* **12**, 5051 (1975).

[He41] K. H. Hellwege, *Ann. Phys.* **40**, 529 (1941).

[He49] K. H. Hellwege, *Ann. Phys.* **4**, 95, 127, 136, 143, 150, 357 (1949).

[He67] J. Heber, *Phys. Kondens. Mater.* **6**, 381 (1967).

[He72] K. H. Hellwege, *Phys. Kondens. Mater.* **15**, 1 (1972).

[He73] K. H. Hellwege, *Phys. Kondens. Mater.* **16**, 1 (1973).

[He76] K. H. Hellwege, "Einführung in die Festkörperphysik." Springer, Berlin and New York, 1976.

[HeCh68] D. E. Henric and G. R. Choppin, *J. Chem. Phys.* **49**, 477 (1968).

[HeHe51] A. M. Hellwege and K. H. Hellwege, *Z. Phys.* **130**, 549 (1951).

[HeHe52] A. M. Hellwege and K. H. Hellwege, *Z. Phys.* **133**, 174 (1952).

[HeHe70] J. Heber, K. H. Hellwege, U. Köbler, and H. Murmann, *Z. Phys.* **237**, 189 (1970).

[HeHe72a] H. Heischmann, K. H. Hellwege, and S. Leutloff, *Phys. Kondens. Mater.* **15**, 10 (1972).

[HeHe72b] H. Heischmann, K. H. Hellwege, and S. Leutloff, *Phys. Kondens. Mater.* **15**, 23 (1972).

[HeHe73] H. Heischmann, K. H. Hellwege, and S. Leutloff, *Phys. Kondens. Mater.* **16**, 9 (1973).
[HeHo61] K. H. Hellwege, G. Horstik, S. Hüfner, and H. Lämmermann, *Z. Phys.* **165**, 253 (1961).
[HeHu73] J. P. Hessler and C. A. Hutchison, Jr., *Phys. Rev. B* **8**, 1822 (1973).
[HeHü60] K. H. Hellwege, S. Hüfner, and H. G. Kahle, *Z. Phys.* **160**, 149 (1960).
[HeHü62] K. H. Hellwege, S. Hüfner, and D. Kuse, *Z. Phys.* **167**, 258 (1962).
[HeHü67] K. H. Hellwege, S. Hüfner, and J. Pelzl, *Z. Phys.* **203**, 227 (1967).
[HeJo57] K. H. Hellwege, U. Johnson, H. G. Kahle, and G. Schaack, *Z. Phys.* **148**, 112 (1957).
[HeMu67] J. R. Henderson, M. Muramoto, and J. B. Gruber, *J. Chem. Phys.* **46**, 2515 (1967).
[HeMu77] J. Heber and H. Murmann, *Z. Phys. B* **26**, 145 (1977).
[HeOr65] K. H. Hellwege, E. Orlich, and G. Schaack, *Phys. Kondens. Mater.* **4**, 196 (1965).
[HiHü70] P. Hill and S. Hüfner, *Z. Phys.* **240**, 168 (1970).
[HiWh66] J. C. Hill and R. G. Wheeler, *Phys. Rev.* **152**, 218 (1966).
[Ho62] S. Van Houten, *Phys. Lett.* **2**, 215 (1962).
[HoLy76] T. Holstein, S. K. Lyo, and R. Orbach, *Phys. Rev. Lett.* **36**, 891 (1976).
[HoSh68] L. Holmes, R. Sherwood, and L. G. Van Uitert, *J. Appl. Phys.* **39**, 1373 (1968).
[HoSo64] R. H. Hoskins and B. H. Soffer, *Appl. Phys. Lett.* **4**, 22 (1964).
[HsPo75] Ch. Hsu and C. Powell, *Phys. Rev. Lett.* **35**, 734 (1975).
[Hu64] M. T. Hutchings, *Solid State Phys.* **16**, 227 (1964).
[HuBi68] M. T. Hutchings, R. J. Birgeneau, and W. P. Wolf, *Phys. Rev.* **168**, 1026 (1968).
[HuLi76] C. A. Hutchison, Jr., and E. D. Lin, *in Proc. 1975 Int. Conf. Luminesc.* North-Holland Publ., Amsterdam, 1976.
[HuRa63] M. T. Hutchings and D. K. Ray, *Proc. Phys. Soc.* **81**, 663 (1963).
[HuWo64] M. T. Hutchings and W. P. Wolf, *J. Chem. Phys.* **41**, 617 (1964).
[Hü62a] S. Hüfner, *Z. Phys.* **168**, 74 (1962).
[Hü62b] S. Hüfner, *Z. Phys.* **169**, 417 (1962).
[Hü71] S. Hüfner, *J. Phys. C1* **32**, 710 (1971).
[Hü72] S. Hüfner, *IEEE Trans. Magn.* **MAG-8**, 151 (1972).
[Hü76] S. Hüfner, *in* "Magnetism Current Topics" (S. Foner, ed.), p. 641. Gordon and Breach, New York, 1976.
[HüHo68] S. Hüfner, L. Holmes, F. Varsanyi, and L. G. Van Uitert, *Phys. Rev.* **171**, 507 (1968).
[ImYe64] G. F. Imbusch, W. M. Yen, A. L. Shawlow, D. E. McCumber, and M. D. Sturge, *Phys. Rev.* **133**, A1029 (1964).
[JaMi76] M. Jacob, H. Micklitz, and K. Luchner, *Phys. Lett.* **57A**, 67 (1976).
[JaNa70] A. Jayaraman, V. Narayanamurti, E. Bucher and R. G. Maines, *Phys. Rev. Lett.* **25**, 368, 1430 (1970).
[JaTe37] H. A. Jahn and E. Teller, *Proc. Roy. Soc.* **A161**, 220 (1937).
[Je63] C. D. Jeffries, "Dynamic Nuclear Orientation." Wiley, New York, 1963.
[Jo63] L. F. Johnson, *J. Appl. Phys.* **34**, 897 (1963).
[Jo66] L. F. Johnson, *in* "Lasers" (A. K. Levine, ed.), Vol. 1, p. 137. Dekker, New York, 1966.
[JoBo62] L. F. Johnson, G. D. Boyd, K. Nassau, and R. R. Soden, *Proc. IRE* (correspondence) **50**, 213 (1962).

[JoGe65] L. F. Johnson, J. E. Geusic, and L. G. Van Uitert, *Appl. Phys. Lett.* **7**, 127 (1965).

[JoJu64] C. K. Jørgensen and B. R. Judd, *Mol. Phys.* **8**, 281 (1964).

[JoPa63] C. K. Jørgensen, R. Pappalardo, and H. H. Schmidtke, *J. Chem. Phys.* **39**, 1422 (1963).

[JoTh63] L. F. Johnson and R. A. Thomas, *Phys. Rev.* **131**, 2038 (1963).

[Ju57] B. R. Judd, *Proc. Roy. Soc.* **A241**, 414 (1957).

[Ju59] B. R. Judd, *Mol. Phys.* **2**, 407 (1959).

[Ju62] B. R. Judd, *Phys. Rev.* **127**, 750 (1962).

[Ju63] B. R. Judd, "Operator Techniques in Atomic Spectroscopy." McGraw-Hill, New York, 1963.

[Ju66a] B. R. Judd, *Phys. Rev.* **141**, 4 (1966).

[Ju66b] B. R. Judd, *J. Chem. Phys.* **44**, 839 (1966).

[Ju77a] B. R. Judd, *J. Chem. Phys.* **66**, 3163 (1977).

[Ju77b] B. R. Judd, *Phys. Rev. Lett.* **39**, 242 (1977).

[JuCr68] B. R. Judd, H. M. Crosswhite, and H. Crosswhite, *Phys. Rev.* **169**, 130 (1968).

[Ka61] H. G. Kahle, *Z. Phys.* **161**, 486 (1961).

[Ka70] A. J. Kassman, *J. Chem. Phys.* **53**, 4118 (1970).

[Ka73] T. Kasuya, *CRC Critical Rev. Solid State Sci.* **3**, 131 (1973).

[KaBe74] A. Kasten and P. J. Becker, *J. Phys. C: Solid State Phys.* **7**, 3120 (1974).

[KaGa61a] W. Kaiser, C. G. B. Garrett, and D. L. Wood, *Phys. Rev.* **123**, 766 (1961).

[KaGa61b] W. Kaiser and C. G. B. Garret, *Phys. Rev. Lett.* **7**, 229 (1961).

[KaHa72a] T. Kambara, W. J. Haas, F. H. Spedding, and R. H. Good, Jr., *J. Chem. Phys.* **56**, 4475 (1972).

[KaHa72b] T. Kambara, W. J. Haas, F. H. Spedding, and R. H. Good, *J. Chem. Phys.* **58**, 672 (1972).

[KaKl71] H. G. Kahle, L. Klein, G. Müller-Vogt, and H. C. Schopper, *Phys. Status Solidi* (*b*) **44**, 619 (1971).

[KaSi74] H. G. Kahle, A. Simon, and W. Wüchner, *Phys. Status Solidi* (*b*) **61**, K53 (1974).

[Ke37] J. A. A. Ketelaar, *Physica* **4**, 619 (1937).

[Ki71] C. Kittel, "Introduction to Solid State Physics," 4th ed., p. 215. Wiley, New York, 1971.

[KiDu62a] Z. J. Kiss and R. C. Duncan, Jr., *Proc. IRE* **50**, 1532 (1962).

[KiDu62b] Z. J. Kiss and R. C. Duncan, Jr., *Proc. IRE* **50**, 1531 (1962).

[KiCr70] J. F. Kielkopf and H. M. Crosswhite, *J. Opt. Soc. Am.* **60**, 347 (1970).

[KiYo64] Z. J. Kiss and P. N. Yokom, *J. Chem. Phys.* **41**, 1511 (1964).

[Ko64] J. A. Koningstein, *Phys. Rev.* **136**, A717 (1964).

[KoDi63] G. F. Koster, J. O. Dimmock, R. G. Wheeler, and H. Statz, "Properties of the Thirty-Two Point Groups." MIT Press, Cambridge, Massachusetts, 1963.

[KoGe64a] J. A. Koningstein and J. E. Geusic, *Phys. Rev.* **136**, A711 (1964).

[KoGe64b] J. A. Koningstein and J. E. Geusic, *Phys. Rev.* **136**, A726 (1964).

[KoMo73] J. A. Koningstein and O. S. Mortensen, *in* "The Raman Effect" (A. Anderson, ed.), p. 519. Dekker, New York, 1973.

[KoScha70] J. A. Koningstein and G. Schaack, *Phys. Rev. B* **2**, 1242 (1970).

[Kr30] H. A. Kramers, *Proc. Amsterdam Acad.* **33**, 959 (1930).

[Kr66] W. F. Krupke, *Phys. Rev.* **145**, 325 (1966).

[Kr71] W. F. Krupke, *IEEE J. Quantum Electron. QE-7*, 153 (1971).

[KrMo73] N. Krasutsky and H. W. Moos, *Phys. Rev. B* **8**, 1010 (1973).

[Ku69] T. Kushida, *Phys. Rev.* **185**, 500 (1969).
[La74] M. Lax, "Symmetry Principles in Solid State and Molecular Physics." Wiley, New York, 1974.
[LaJe66] G. H. Larson and C. D. Jeffries, *Phys. Rev.* **141**, 461 (1966).
[LaKa75] J. Laugsch, H. G. Kahle, M. Schwab, and W. Wüchner, *Physica* **80B**, 269 (1975).
[LaKe71] D. P. Landau, B. E. Keen, B. Schneider, and W. P. Wolf, *Phys. Rev. B* **3**, 2310 (1971).
[LaLo77] C. B. Layne, W. H. Lowdermilk, and M. J. Weber, *Phys. Rev. B* **16**, 10 (1977).
[Le64] P. M. Levy, *Phys. Rev.* **135**, A155 (1964).
[Le66a] P. M. Levy, *Phys. Rev.* **147**, 311 (1966).
[Le66b] A. K. Levine (ed.), "Lasers." Dekker, New York, 1966, Vol. 1; Vol. 2, 1968.
[Le68] M. J. M. Leask, *J. Appl. Phys.* **39**, 908 (1968).
[Le69] P. M. Levy, *Phys. Rev.* **177**, 509 (1969).
[LiBu75] H. G. Lipson, J. R. Buckmelter, and C. O. Dugger, *J. Non-Cryst. Solids* **17**, 27 (1975).
[Lo67] E. Loh, *Phys. Rev.* **158**, 273 (1967) and *Phys. Rev.* **140**, A1463 (1965).
[Ma60] T. H. Maiman, *Nature (London)* **187**, 493 (1960).
[Ma61] J. S. Margolis, *J. Chem. Phys.* **35**, 1367 (1961).
[Ma63a] P. B. Mauer, *Appl. Opt.* **3**, 153 (1963).
[Ma63b] R. D. Maurer, *Appl. Opt.* **2**, 87 (1963).
[Ma66] A. A. Maradudin, *Solid State Phys.* **18**, 381 (1966).
[Ma68] S. L. Marshall (ed.), "Laser Technology and Applications." McGraw-Hill, New York, 1968.
[MaBo61] B. T. Matthias, R. M. Bozorth, and J. H. Van Vleck, *Phys. Rev. Lett.* **7**, 160 (1961).
[MaDeSh70] M. M. Mann and L. G. DeShazer, *J. Appl. Phys.* **41**, 2951 (1970).
[MaOr66] A. A. Manenkov and R. Orbach (eds.), "Spin Lattice Relaxation in Ionic Solids," Harper and Row, New York, 1966.
[McCl59] D. S. McClure, *Solid State Phys.* **9**, 399 (1959).
[McClKi63] D. S. McClure and Z. J. Kiss, *J. Chem. Phys.* **39**, 3251 (1963).
[McCuStu63] D. E. McCumber and M. D. Sturge, *J. Appl. Phys.* **34**, 1682 (1963).
[McGuAr62] T. R. McGuire, B. E. Argyle, M. W. Shafer, and J. S. Smart, *Appl. Phys. Lett.* **1**, 17 (1962).
[Me65] S. Methfessel, *Z. Angew. Phys.* **18**, 414 (1965).
[Me76] R. S. Meltzer, *Solid State Commun.* **20**, 553 (1976).
[MeCo68] R. S. Meltzer, E. Cohen, and H. W. Moos, *Phys. Rev. Lett.* **21**, 1690 (1968).
[MeMa68] S. Methfessel and D. Matthis, *Springer Tracts Phys.* **18**, 389 (1968).
[MeMo70] R. S. Meltzer and H. W. Moos, *J. Appl. Phys.* **41**, 1240 (1970).
[MeMo72] R. S. Meltzer and H. W. Moos, *Phys. Rev. B* **6**, 264 (1972).
[MePy73] R. L. Melcher, E. Pytte, and B. A. Scott, *Phys. Rev. Lett.* **31**, 307 (1973).
[MiDe70] T. Miyakawa and D. L. Dexter, *Phys. Rev. B* **1**, 2961 (1970).
[MoWo76] C. A. Morrison, D. E. Wortmann, and N. Karayianis, *J. Phys. C Solid State Phys.* **9**, L191 (1976).
[Mö58] R. L. Mößbauer, *Z. Phys.* **151**, 124 (1958).
[MöBa64] R. L. Mößbauer, R. G. Barnes, E. Kankeleit, and J. M. Poindexter, *Phys. Rev.* **136**, A175 (1964).

[MuHa67] T. Murao, W. J. Haas, R. W. G. Syme, and F. H. Spedding, *J. Chem. Phys.* **47**, 1572 (1967).
[MuSp65] T. Murao, F. H. Spedding, and R. H. Good, Jr., *J. Chem. Phys.* **42**, 993 (1965).
[NaShi70] E. Nakazawa and S. Shionoya, *Phys. Rev. Lett.* **25**, 1710 (1970).
[Ne71] D. J. Newman, *Adv. Phys.* **20**, 197 (1971).
[Ne74] V. Nekvasil, *J. Phys. C Solid State Phys.* **7**, L246 (1974).
[Ne77a] D. J. Newman, *J. Phys. C, Solid State Phys.* **10**, 4753 (1977).
[Ne77b] D. J. Newman, *J. Phys. C, Solid State Phys.* **10**, L617 (1977).
[NePa64] L. Néel, R. Pauthenet, and B. Dreyfus, *Progr. Low Temp. Phys.* **4**, 344 (1964).
[NeSte69] D. J. Newman and G. F. Stedman, *J. Chem. Phys.* **51**, 3013 (1969).
[NeSte71] D. J. Newman and G. E. Stedman, *J. Phys. Chem. Solids* **32**, 535 (1971).
[NiKo64] C. W. Nielson and G. F. Koster, "Spectroscopic Coefficients for p^n, d^n, and f^n Configurations." MIT Press, Cambridge, Massachusetts, 1964.
[Of62] G. S. Ofelt, *J. Chem. Phys.* **37**, 511 (1962).
[OfRa65] S. Ofer, M. Rakavy, E. Segal, and B. Khurgin, *Phys. Rev.* **138**, A241 (1965).
[OlEl60] M. C. Olsen and B. Elbek, *Nucl. Phys.* **15**, 134 (1960).
[Or61] R. Orbach, *Proc. Roy. Soc.* **A264**, 458, 485 (1961).
[Or67] R. Orbach, *in Conf. Properties of Ions in Crystals.* p. 445. Wiley (Interscience), New York, 1967.
[Or75] R. Orbach, *in* "Optical Properties of Ions in Solids" (B. Di Bartolo, ed.), p. 355. Plenum Press, New York, 1975.
[OrHü69] E. Orlich and S. Hüfner, *J. Appl. Phys.* **40**, 1503 (1969).
[OrHü70] E. Orlich and S. Hüfner, *Z. Phys.* **232**, 418 (1970).
[OsZi71] F. W. Ostermayer, Jr., J. P. van der Ziel, H. M. Marcos, and J. E. Geusic, *Phys. Rev. B* **3**, 2698 (1971).
[Pa58] R. Pauthenet, *Ann. Phys. (Paris)* **3**, 424 (1958).
[PaDi76] L. Passell, O. W. Dietrich, and J. Als-Nielsen, *Phys. Rev. B* **14**, 4897 (1976).
[Pe70] J. Pelzl, Thesis, Darmstadt (1970).
[Pe75] R. D. Peacock, *Structure and Bonding*, **22**, 83 (1975).
[PePe77] N. Pelletier-Allard, R. Pelletier, and C. Delsart, *J. Phys. C, Solid State Phys.* **10**, 2005 (1977).
[PeHü70] J. Pelzl, S. Hüfner, and S. Scheller, *Z. Phys.* **231**, 377 (1970).
[PeSte78] R. D. Peacock and B. Stewart, *Chem. Phys. Lett*, **55**, 295 (1978).
[PiDi67] A. H. Piksis, G. H. Dieke, and H. M. Crosswhite, *J. Chem. Phys.* **47**, 5083 (1967).
[PoOr61] M. J. D. Powell and R. Orbach, *Proc. Phys. Soc.* **78**, 753 (1961).
[Pr66] G. A. Prinz, *Phys. Lett.* **20**, 323 (1966); *Phys. Rev.* **152**, 474 (1966).
[Pr71] R. J. Pressley (ed.), "Handbook of Lasers." CRC Press, Cleveland, 1971.
[PrCo68] G. A. Prinz and E. Cohen, *Phys. Rev.* **165**, 335 (1968).
[Ra42] G. Racah, *Phys. Rev.* **62**, 438 (1942).
[Ra43] G. Racah, *Phys. Rev.* **63**, 367 (1943).
[Ra49] G. Racah, *Phys. Rev.* **76**, 1352 (1949).
[Ra66] R. C. Rau, *Acta Crystallogr.* **20**, 716 (1966).
[RaKr67] K. Rajnak and W. F. Krupke, *J. Chem. Phys.* **46**, 3532 (1967).
[RaKr68] K. Rajnak and W. F. Krupke, *J. Chem. Phys.* **48**, 3343 (1968).
[RaRa67] A. K. Raychaudhuri and D. K. Ray, *Proc. Phys. Soc.* **90**, 839 (1967).
[RaWy63] K. Rajnak and B. G. Wybourne, *Phys. Rev.* **132**, 280 (1963).

[Re73] R. Reisfeld, *Structure and Bonding* **13**, 53 (1973).
[Re75] R. Reisfeld, *Structure and Bonding* **22**, 123 (1975).
[Re76] R. Reisfeld, *Structure and Bonding* **30**, 65 (1976).
[ReGl64] R. Reisfeld and A. Glasner, *J. Opt. Soc. Am.* **54**, 331 (1964).
[ReJo78] R. Reisfeld and C. K. Jørgensen, *Inorg. Chem. Concepts* **1** (1978).
[RiDeSh69] D. K. Rice and L. G. DeShazer, *Phys. Rev.* **186**, 387 (1969).
[RiMo67] L. A. Riseberg and H. W. Moss, *Phys. Rev. Lett.* **19**, 1423 (1967).
[RiMo68] L. A. Riseberg and H. W. Moos, *Phys. Rev.* **174**, 429 (1968).
[RiSa63] I. Richman, R. A. Satten, and E. Y. Wong, *J. Chem. Phys.* **39**, 1833 (1963).
[RiWe76] L. A. Riseberg and M. J. Weber, Relaxation phenomena in rare earth luminescence, *in* "Progress in Optics" (E. Wolf, ed.), Vol. XIV. North-Holland Publ., Amsterdam, 1976.
[Ro57] M. E. Rose, "Elementary Theory of Angular Momentum." Wiley, New York, 1957.
[RoBi59] M. Rotenberg, R. Bivins, N. Metropolis, and J. K. Wooten, "The 3-*j* and 6-*j* Symbols." MIT Press, Cambridge, Massachusetts, 1959.
[Sa64] R. A. Satten, *J. Chem. Phys.* **40**, 1200 (1964).
[SchäJo65] C. E. Schäffer and C. K. Jørgensen, *Mol. Phys.* **9**, 401 (1965).
[Schoe75] J. Schoenes: Z. Physik **B20**, 345 (1975).
[SchwBe74] M. Schwab, W. Berkhahn, H. G. Kahle, A. Simon, and W. Wüchner, *Phys. Status Solidi* (*b*) **65**, 613 (1974).
[ScJe62] P. L. Scott and C. D. Jeffries, *Phys. Rev.* **127**, 32 (1962).
[ScTo58] A. L. Schawlow and C. H. Townes, *Phys. Rev.* **112**, 1940 (1958).
[SeAr70] D. Sengupta and J. O. Artman, *Phys. Rev. B* **1**, 2986 (1970).
[ShiNa76] S. Shionoya, S. Nagakura, and S. Sugano, *Proc. 1975 Int. Conf. Luminesc.* North-Holland Publ., Amsterdam, 1976.
[SiFi64] R. H. Silsbee and D. B. Fitchen, *Rev. Mod. Phys.* **36**, 432 (1964).
[Sm68] R. Smitz, *IEEE J. Quantum Electron.* **4**, 505 (1968).
[SmUn65] K. F. Smith and P. J. Unsworth, *Proc. Phys. Soc.* **86**, 1249 (1965).
[SmTh66] D. Smith and J. H. M. Thornley, *Proc. Phys. Soc.* **89**, 779 (1966).
[Sn61] E. Snitzer, *Phys. Rev. Lett.* **7**, 444 (1961); see also E. Snitzer and G. Young, *in* "Lasers" (A. K. Levine, ed.), Vol. 2, p. 191. Dekker, New York, 1968.
[SnWo65] E. Snitzer and R. Woodcock, *Appl. Phys. Lett.* **6**, 45 (1965); see also E. Snitzer and G. Young, *in* "Lasers" (A. K. Levine, ed), Vol. 2, p. 191. Dekker, New York, 1968.
[SnYo68] E. Snitzer and C. G. Young, *in* "Lasers" (A. K. Levine, ed.), Vol. 2, p. 191. Dekker, New York, 1968.
[SoHo65] B. H. Soffer and R. H. Hoskins, *Appl. Phys. Lett.* **6**, 200 (1965).
[SoMu63] R. Solomon and L. Mueller, *Appl. Phys. Lett.* **3**, 135 (1963).
[StaMa76] R. G. Stafford, H. Masui, R. L. Farrow, and R. K. Chang, *J. Appl. Phys.* **47**, 2483 (1976).
[Ste51] R. M. Sternheimer, *Phys. Rev.* **84**, 244 (1951).
[Ste52] K. W. H. Stevens, *Proc. Phys. Soc.* **A65**, 209 (1952).
[Ste67] K. W. H. Stevens, *Rep. Progr. Phys.* **30**, 189 (1967).
[Sti78] C. M. Stickley, *Phys. Today*, May, 50 (1978).
[Sto69] A. M. Stoneham, *Rev. Mod. Phys.* **41**, 82 (1969).
[Su65] J. Sugar, *Phys. Rev. Lett.* **14**, 731 (1965).
[SuTa70] S. Sugano, Y. Tanabe, and H. Kamimura, "Multiplets of Transition Metal Ions in Crystals." Academic Press, New York, 1970.

[SyHa68] R. W. G. Syme, W. J. Haas, F. H. Spedding, and R. H. Good, Jr., *J. Chem. Phys.* **48**, 2772 (1968).

[Sz70] A. Szabo, *Phys. Rev. Lett.* **25**, 924 (1970).

[Ta72] K. N. R. Taylor and N. I. Darby, "Physics of Rare Earth Solids." Chapman and Hall, London, 1972.

[Ta76] N. Takeuchi, *J. Luminesc.* **12/13**, 743 (1976).

[TaSz74] N. Takeuchi and A. Szabo, *Phys. Lett.* **50A**, 361 (1974).

[ThSi63] K. S. Thomas, S. Singh, and G. H. Dieke, *J. Chem. Phys.* **38**, 2180 (1963).

[Ti64] M. Tinkham, "Group Theory and Quantum Mechanics." McGraw–Hill, New York, 1964.

[TiBu67] T. Timusk and M. Buchanan, *Phys. Rev.* **164**, 345 (1967).

[To53] C. H. Townes, *J. Inst. Elec. Commun. Eng.* (*Jpn*) **36**, 650 (1953).

[ToWe75] B. C. Tofield, H. P. Weber, T. C. Daman, and P. F. Liao, *J. Solid State Chem.* **12**, 207 (1975).

[Tr63] E. D. Trifanov, *Sov. Phys. Dokl.* **7**, 1105 (1963).

[Ur78] W. Urland, *Chem. Phys. Lett.* **53**, 296 (1978).

[Va71] F. Varsanyi, *Appl. Phys. Lett.* **19**, 169 (1971).

[Va76] C. M. Varma, *Rev. Mod. Phys.* **48**, 219 (1976).

[VaDi61] F. Varsanyi and G. H. Dieke, *Phys. Rev. Lett.* **7**, 442 (1961).

[VeAy70] C. Veyret, J. B. Ayasse, J. Chaussy, J. Mareschal, and J. Sivardière, *J. Phys.* **31**, 607 (1970).

[vVl37] J. H. van Vleck, *J. Phys. Chem.* **41**, 67 (1937).

[vVl62] J. H. van Vleck, *Rev. Mat. Fis. Teor. Univ. Nat. Tucuman* **14**, 189 (1962).

[vVl66] J. H. van Vleck, *J. Phys. Chem. Solids* **27**, 1047 (1966).

[Wa65] M. Wagner, *Phys. Kondens. Mat.* **4**, 71 (1965).

[Wa71] P. Wachter, *Z. Angew. Phys.* **32**, 171 (1971).

[Wa73] P. Wachter, *CRC Critical Rev. Solid State Sci.* **3**, 189 (1973).

[Wa74a] R. K. Watts, *in* "Optical properties of Ions in Solids" (B. DiBartolo, ed.), p. 307. Plenum Press, New York, 1975.

[Wa74b] R. K. Watts, *in* "Optical Properties of Ions in Solids" (B. DiBartolo, ed.), p. 337. Plenum Press, New York, 1975.

[Wa77] P. Wachter, *in* "Handbook of the Physics and Chemistry of the Rare Earths" (K. A. Gschneider and L. Eyring, eds.), Vol. 1, Chapter 19. North-Holland Publ., Amsterdam, 1977.

[WaBr65a] M. Wagner and W. E. Bron, *Phys. Rev.* **139**, A223, (1965).

[WaBr65b] M. Wagner and W. E. Bron, *Phys. Rev.* **139**, A233 (1965).

[WaFr64a] R. E. Watson and A. J. Freeman, *Phys. Rev.* **135**, A1209 (1964).

[WaFr64b] R. E. Watson and A. J. Freeman, *Phys. Rev.* **133**, A1571 (1964).

[WaFr64c] R. E. Watson and A. J. Freeman, *Phys. Rev.* **134**, A1526 (1964).

[WaFr67] R. E. Watson and A. J. Freeman, *Phys. Rev.* **156**, 251 (1967).

[WaLe71] R. L. Wadsack, J. L. Lewis, B. E. Argyle, and R. K. Chang, *Phys. Rev. B* **3**, 4342 (1971).

[WaRi72] R. K. Watts and H. J. Richter, *Phys. Rev. B* **6**, 1584 (1972).

[We68] M. J. Weber, *Phys. Rev.* **171**, 283 (1968).

[We71] M. J. Welm, E. J. Sharp, and J. E. Müller, *J. Phys. Chem. Solids* **32**, 2275 (1971).

[WeCo71] G. K. Wertheim, R. L. Cohen, A. Rosencwaig, and H. J. Guggenheim, *Phys. Rev. Lett.* **27**, 505 (1971).

[WeDa73] H. P. Weber, T. L. Damen, H. G. Danielmeyer, and B. C. Tofield, *Appl. Phys. Lett.* **22**, 534 (1973).

[WeMa72] M. J. Welm, B. H. Matsinger, V. L. Doulan, and G. T. Suvratt, *J. Chem. Phys.* **57**, 562 (1972).

[WeRe68] R. G. Wheeler, F. M. Reames, and E. J. Wachtel, *J. Appl. Phys.* **39**, 915 (1968).

[WeVa71] M. J. Weber and T. E. Varitimos, *J. Appl. Phys.* **42**, 4996 (1971).

[WeVa73] M. J. Weber, T. E. Varitimos, and B. H. Matsinger, *Phys. Rev. B* **8**, 47 (1973).

[WeWo72] M. R. Wells and R. D. Worswick, *Phys. Lett.* **A42**, 269 (1972).

[Wh69] R. L. White, *J. Appl. Phys.* **40**, 1061 (1969).

[Wi61] K. A. Wickersheim, *Phys. Rev.* **122**, 1376 (1961).

[WiNo66] H. H. Wickman and I. Nowik, *Phys. Rev.* **142**, 115 (1966).

[WiWh60] K. A. Wickersheim and R. L. White, *Phys. Rev. Lett.* **4**, 123 (1960).

[WiWh62] K. A. Wickersheim and R. L. White, *Phys. Rev. Lett.* **8**, 483 (1962).

[Wo65] W. P. Wolf, *in Proc. Int. Conf. Magn., Nottingham, 1964*. The Inst. of Physics and The Physical Society, London, 1965.

[Wo71] W. P. Wolf, *J. Physi.* **32**, Cl-26 (1971).

[WoBa62] W. P. Wolf, M. Ball, M. T. Hutchings, M. J. M. Leask, and A. F. G. Wyatt, *J. Phys. Soc. Jpn. Suppl. B-1* **17**, 443 (1962).

[WoCo73] D. K. Wohlleben and B. R. Coles, *in* "Magnetism" (H. Suhl, ed.), Vol. V, p. 3. Academic Press, New York, 1973.

[WoKa62] D. L. Wood and W. Kaiser, *Phys. Rev.* **126**, 2079 (1962).

[WoRi61] E. Y. Wong and I. Richman, *J. Chem. Phys.* **34**, 1182 (1961).

[Wr76] J. C. Wright, *in* "Topics in Applied Physics" (F. K. Fong, ed.), p. 239. Springer–Verlag, Berlin and New York, 1976.

[WrMo71a] J. C. Wright, H. W. Moos, J. H. Colwell, B. W. Mangum, and D. D. Thornton, *Phys. Rev. B* **3**, 843 (1971).

[WrMo71b] J. C. Wright and H. W. Moos, *Phys. Rev. B* **4**, 163 (1971).

[WüBö72] W. Wüchner, W. Böhm, H. G. Kahle, A. Kasten, and J. Laugsch, *Phys. Status Solidi* (*b*) **54**, 273 (1972).

[WüLa73] W. Wüchner and J. Laugsch, *Int. J. Magn.* **5**, 181 (1973).

[Wy65] B. G. Wybourne, "Spectroscopic Properties of Rare Earths." Wiley, New York, 1965.

[Wy66] B. G. Wybourne, *Phys. Rev.* **148**, 317 (1966).

[Wy68] A. F. G. Wyatt, *J. Phys. C* **1**, 684 (1968).

[Ya62] A. Yariv, *Proc. IRE* **50**, 1699 (1962).

[YaGo63] A. Yariv and J. P. Gordon, *Proc. IEEE* **51**, 4 (1963).

[YaOt69] H. Yamamoto, Y. Otomo, and T. Kano, *J. Phys. Soc. Jpn.* **26**, 137 (1969).

[YaPo62] A. Yariv, S. P. S. Porto, and K. Nassau, *J. Appl. Phys.* **33**, 2519 (1962).

[YaShi72] N. Yamada, S. Shionoya, and T. Kushida, *J. Phys. Soc. Jpn.* **32**, 1577 (1972).

[YeSc64] W. M. Yen, W. C. Scott, and A. L. Schawlow, *Phys. Rev.* **136**, A271 (1964).

[Yo63] C. G. Young, *Appl. Phys. Lett*, **2**, 151 (1963).

[Zi54] J. M. Ziman, *Proc. Roy. Soc.* (*London*) **A226**, 436 (1954).

[ZiUi69] J. P. van der Ziel and L. G. Van Uitert, *Phys. Rev.* **180**, 343 (1969).

[ZiUi73] J. P. van der Ziel and L. G. Van Uitert, *Phys. Rev. B* **8**, 1889 (1973).

Index